NCES
National Center for
Education Statistics

U.S. Department of Education
Institute of Education Sciences
NCES 2003-013

Teaching Mathematics in Seven Countries

Results From the TIMSS 1999 Video Study

March 2003

James Hiebert
Ronald Gallimore
Helen Garnier
Karen Bogard Givvin
Hilary Hollingsworth
Jennifer Jacobs
Angel Miu-Ying Chui
Diana Wearne
Margaret Smith
Nicole Kersting
Alfred Manaster
Ellen Tseng
Wallace Etterbeek
Carl Manaster
Patrick Gonzales
James Stigler

For sale by the Superintendent of Documents, U.S. Government Printing Office
Internet: bookstore.gpo.gov Phone: toll free (866) 512-1800; DC area (202) 512-1800
Fax: (202) 512-2250 Mail: Stop SSOP, Washington, DC 20402-0001
ISBN 0-16-051381-2

U.S. Department of Education
Rod Paige, *Secretary*

Institute of Education Sciences
Grover J. Whitehurst, *Director*

National Center for Education Statistics
Val Plisko, *Associate Commissioner*

The National Center for Education Statistics (NCES) is the primary federal entity for collecting, analyzing, and reporting data related to education in the United States and other nations. It fulfills a congressional mandate to collect, collate, analyze, and report full and complete statistics on the condition of education in the United States; conduct and publish reports and specialized analyses of the meaning and significance of such statistics; assist state and local education agencies in improving their statistical systems; and review and report on education activities in foreign countries.

NCES activities are designed to address high priority education data needs; provide consistent, reliable, complete, and accurate indicators of education status and trends; and report timely, useful, and high quality data to the U.S. Department of Education, the Congress, the states, other education policymakers, practitioners, data users, and the general public.

We strive to make our products available in a variety of formats and in language that is appropriate to a variety of audiences. You, as our customer, are the best judge of our success in communicating information effectively. If you have any comments or suggestions about this or any other NCES product or report, we would like to hear from you. Please direct your comments to:

National Center for Education Statistics
Institute of Education Sciences
U.S. Department of Education
1990 K Street, NW
Washington, DC 20006

March 2003

The NCES World Wide Web Home Page is: *http://nces.ed.gov*
The NCES World Wide Electronic Catalog is: *http://nces.ed.gov/pubsearch*

Suggested Citation
U.S. Department of Education, National Center for Education Statistics. *Teaching Mathematics in Seven Countries: Results From the TIMSS 1999 Video Study,* NCES (2003-013), by James Hiebert, Ronald Gallimore, Helen Garnier, Karen Bogard Givvin, Hilary Hollingsworth, Jennifer Jacobs, Angel Miu-Ying Chui, Diana Wearne, Margaret Smith, Nicole Kersting, Alfred Manaster, Ellen Tseng, Wallace Etterbeek, Carl Manaster, Patrick Gonzales, and James Stigler. Washington, DC: 2003.

For ordering information on this report, write:
U.S. Department of Education, ED Pubs
P.O. Box 1398
Jessup, MD 20794-1398

or call toll free 1-877-4ED-PUBS or go to the Internet: *http://nces.ed.gov/timss*

Content Contact:
TIMSS Customer Service: (202) 502-7421 fax: (202) 502-7455
Email: timss@ed.gov

Front cover image: Harnett/Hanzon/Getty Images.

ACKNOWLEDGMENTS

The authors of this report wish to recognize the many people who contributed to making this report possible, from its conceptualization to data collection to analysis to writing. In addition to the many contributors listed in appendix B of this report, the authors wish to thank Ilona Berkovits, Shelley Burns, Arnold Goldstein, William Hussar, Val Plisko, Eugene Owen, and Marilyn Seastrom of NCES, Christopher Calsyn, Erin Pahlke, and David Miller of the Education Statistics Services Institute, Margaret Cozzens, Colorado Institute of Technology, Joanne Bogart, El Paso Collaborative for Academic Excellence, and Patsy Wang-Iverson, Research for Better Schools for their helpful technical and editorial feedback.

Perhaps most importantly, the authors wish to thank the hundreds of teachers and thousands of students who allowed the camera into the classroom and their lives. This study would have been impossible without their willingness to open the classroom door.

The TIMSS 1999 Video Study was conducted by LessonLab, Inc. under contract to the National Center for Education Statistics, U.S. Department of Education. The U.S. National Science Foundation and the participating countries provided additional funding for the study.

TABLE OF CONTENTS

ACKNOWLEDGMENTS iii

LIST OF TABLES AND FIGURES vii

CHAPTER 1 | Introduction 1
    Studying Classroom Teaching Across Countries 2
    Building on the TIMSS 1995 Video Study 9
    What Can Be Found in This Report? 11
    A Lens Through Which to View This Report 13

CHAPTER 2 | Context of the Lessons 15
    The Teachers 15
    Teachers and Current Ideas About Teaching and Learning Mathematics 24
    Teachers' Perceptions of the Typicality of the Videotaped Lesson 26
    Summary 33

CHAPTER 3 | The Structure of Lessons 35
    The Length of Lessons 36
    The Amount of Time Spent Studying Mathematics 38
    The Role of Mathematical Problems 41
    The Purpose of Different Lessons Segments 49
    Public and Private Classroom Interaction 53
    The Role of Homework 56
    Pedagogical Features That Influence Lesson Clarity and Flow 59
    Summary 64

CHAPTER 4 | The Mathematical Content of Lessons 67
    Mathematical Topics Covered During the Lessons 68
    Level of Mathematics Evident in the Lessons 69
    Type of Mathematics Evident in the Lessons 70
    How Mathematics Is Related Over the Lesson 76
    Summary 80

CHAPTER 5 | Instructional Practices: How Mathematics Was Worked On 83

How Mathematical Problems Were Presented and Worked On 83
The Nature of Non-Problem Segments 106
Opportunities to Talk 107
Resources Used During the Lesson 113
Summary 116

CHAPTER 6 | Similarities and Differences in Eighth-Grade Mathematics Teaching Across Seven Countries 119

Locating Similarities and Differences in Eighth-Grade Mathematics Teaching 120
What Are the Relationships Among Features of Mathematics Teaching in Each Country? 123
The Roles Played by Individual Lesson Features Within Different Teaching Systems 148
Conclusions 149

REFERENCES 153

APPENDIX A | Sampling, Questionnaires, Video Data Coding Teams, and Statistical Analyses 161

APPENDIX B | Participants in the TIMSS 1999 Video Study of Mathematics Teaching 173

APPENDIX C | Standard Errors for Estimates Shown in Figures and Tables 179

APPENDIX D | Results From the Mathematics Quality Analysis Group 189

APPENDIX E | Hypothesized Country Models 203

APPENDIX F | Numeric Values for the Lesson Signatures 215

LIST OF TABLES AND FIGURES

TABLE 1.1. TIMSS 1999 Video Study participating countries and their average score on TIMSS 1995 and TIMSS 1999 mathematics assessments 11

TABLE 2.1. Percentage of eighth-grade mathematics lessons taught by teachers who identified one or more major fields of undergraduate and graduate study, by country: 1999 17

TABLE 2.2. Percentage of eighth-grade mathematics lessons taught by teachers certified in various subject areas, by grade level of certification and by country: 1999 18

TABLE 2.3. Mean, median, and range of number of years that teachers reported teaching in general and teaching mathematics, by country: 1999 19

TABLE 2.4. Average hours per week that teachers reported spending on teaching and other school-related activities, by country: 1999 20

FIGURE 2.1. Percentage of eighth-grade mathematics lessons taught by teachers who identified content, process, or perspective goals for the videotaped lesson, by country: 1999 21

TABLE 2.5. Percentage of eighth-grade mathematics lessons taught by teachers who identified specific process goals for the videotaped lesson, by country: 1999 22

TABLE 2.6. Percentage of eighth-grade mathematics lessons taught by teachers who reported that various factors played a "major role" in their decision to teach the content in the videotaped lesson, by country: 1999 23

FIGURE 2.2. Percentage distribution of eighth-grade mathematics lessons according to whether the teachers believed that they were familiar with current ideas in mathematics teaching and learning, by country: 1999 25

FIGURE 2.3. Percentage distribution of eighth-grade mathematics lessons taught by teachers who rated the extent to which the videotaped lesson was in accord with current ideas about teaching and learning mathematics, by country: 1999 26

FIGURE 2.4. Percentage distribution of eighth-grade mathematics lessons taught by teachers who rated how often they used the teaching methods in the videotaped lesson, by country: 1999 28

FIGURE 2.5.	Percentage distribution of eighth-grade mathematics lessons by teachers' ratings of their students' behavior in the videotaped lesson, by country: 1999	29
FIGURE 2.6.	Percentage distribution of eighth-grade mathematics lessons by teachers' ratings of the difficulty of the lesson content compared to usual, by country: 1999	30
FIGURE 2.7.	Percentage distribution of eighth-grade mathematics lessons taught by teachers who rated the influence of the camera on their teaching of the videotaped lesson, by country: 1999	31
FIGURE 2.8.	Length of time averaged over eighth-grade mathematics lessons that teachers reported planning for the videotaped lesson and for similar mathematics lessons, by country: 1999	32
TABLE 2.7.	Average number of eighth-grade mathematics lessons in unit and placement of the videotaped lesson in unit, by country: 1999	33
TABLE 3.1.	Mean, median, range, and standard deviation (in minutes) of the duration of eighth-grade mathematics lessons, by country: 1999	37
FIGURE 3.1.	Box and whisker plots showing the distribution of eighth-grade mathematics lesson durations, by country: 1999	38
FIGURE 3.2.	Average percentage of eighth-grade mathematics lesson time devoted to mathematical work, mathematical organization, and non-mathematical work, by country: 1999	39
TABLE 3.2.	Estimated median time spent in mathematical work per week and per year in eighth grade, by country: 1999	40
FIGURE 3.3.	Average percentage of eighth-grade mathematics lesson time devoted to problem and non-problem segments, by country: 1999	42
TABLE 3.3.	Average number of independent and answered-only problems per eighth-grade mathematics lesson, by country: 1999	44
FIGURE 3.4.	Average percentage of eighth-grade mathematics lesson time devoted to independent problems, concurrent problems, and answered-only problems, by country: 1999	45
FIGURE 3.5.	Average time per independent problem per eighth-grade mathematics lesson (in minutes), by country: 1999	46
FIGURE 3.6.	Box and whisker plots showing the distribution of eighth-grade mathematics lessons based on average length of independent problems, by country: 1999	47
FIGURE 3.7.	Average percentage of independent and concurrent problems per eighth-grade mathematics lesson that were worked on for longer than 45 seconds, by country: 1999	48

FIGURE 3.8.	Average percentage of eighth-grade mathematics lesson time devoted to various purposes, by country: 1999	50
TABLE 3.4.	Percentage of eighth-grade mathematics lessons with at least one segment of each purpose type, by country: 1999	51
FIGURE 3.9.	Percentage of eighth-grade mathematics lessons that were entirely review, by country: 1999	52
TABLE 3.5.	Average number of shifts in purpose per eighth-grade mathematics lesson, by country: 1999	53
TABLE 3.6.	Average percentage of eighth-grade mathematics lesson time devoted to public interaction, private interaction and optional, student presents information, by country: 1999	54
FIGURE 3.10.	Average percentage of private interaction time per eighth-grade mathematics lesson that students spent working individually or in pairs and groups, by country: 1999	55
TABLE 3.7.	Average number of classroom interaction shifts per eighth-grade mathematics lesson, by country: 1999	56
FIGURE 3.11.	Percentage of eighth-grade mathematics lessons in which homework was assigned, by country: 1999	57
TABLE 3.8.	Average number of eighth-grade mathematics problems per lesson assigned as homework and begun in the lesson, and estimated average time per lesson spent on these problems, by country: 1999	58
TABLE 3.9.	Average number of eighth-grade mathematics problems per lesson previously assigned as homework and corrected or discussed during the lesson, and estimated average time per lesson spent on these problems, by country: 1999	59
FIGURE 3.12.	Percentage of eighth-grade mathematics lessons that contained at least one goal statement, by country: 1999	60
FIGURE 3.13.	Percentage of eighth-grade mathematics lessons that contained at least one summary statement, by country: 1999	61
FIGURE 3.14.	Percentage of eighth-grade mathematics lessons with at least one outside interruption, by country: 1999	62
FIGURE 3.15.	Percentage of eighth-grade mathematics lessons with at least one non-mathematical segment at least 30 seconds in length within the mathematics portion of the lesson, by country: 1999	63
FIGURE 3.16.	Percentage of eighth-grade mathematics lessons with at least one public announcement by the teacher during private work time unrelated to the current assignment, by country: 1999	64

TABLE 4.1.	Average percentage of problems per eighth-grade mathematics lesson within each major category and sub-category topic area, by country: 1999	69
FIGURE 4.1.	Average percentage of eighth-grade mathematics problems per lesson at each level of procedural complexity, by country: 1999	71
FIGURE 4.2.	Average percentage of two-dimensional geometry problems at each level of procedural complexity per eighth-grade mathematics lesson in sub-sample of lessons containing two-dimensional geometry problems, by country: 1999	72
FIGURE 4.3.	Average percentage of problems per eighth-grade mathematics lesson that included proofs, by country: 1999	74
FIGURE 4.4.	Percentage of eighth-grade mathematics lessons that contained at least one proof, by country: 1999	74
FIGURE 4.5.	Average percentage of two-dimensional geometry problems that included proofs per eighth-grade mathematics lesson in sub-sample of lessons containing two-dimensional geometry problems, by country: 1999	75
FIGURE 4.6.	Average percentage of eighth-grade mathematics problems (excluding the first problem) per lesson related to previous problems, by country: 1999	77
TABLE 4.2.	Average number of unrelated problems per eighth-grade mathematics lesson, by country: 1999	78
FIGURE 4.7.	Average percentage of two-dimensional geometry problems (excluding the first problem) related to previous problems per eighth-grade mathematics lesson in sub-sample of lessons containing two-dimensional geometry problems, by country: 1999	79
FIGURE 4.8.	Percentage of eighth-grade mathematics lessons that contained problems related to a single topic, by country: 1999	80
FIGURE 5.1.	Average percentage of problems per eighth-grade mathematics lesson that were either set up with the use of a real-life connection, or set up using mathematical language or symbols only, by country: 1999	85
FIGURE 5.2.	Average percentage of problems per eighth-grade mathematics lesson that contained a drawing/diagram, table, and/or graph, by country: 1999	86
FIGURE 5.3.	Average percentage of problems per eighth-grade mathematics lesson that involved the use of physical materials, by country: 1999	88

FIGURE 5.4. Average percentage of two-dimensional geometry problems per eighth-grade mathematics lesson that involved the use of physical materials, by country: 1999 89

FIGURE 5.5. Example of an application problem: "Find the measure of angle x." .90

FIGURE 5.6. Average percentage problems per eighth-grade mathematics lesson that were applications, by country: 1999 91

FIGURE 5.7. Average percentage of problems per eighth-grade mathematics lesson for which a solution was presented publicly in the videotaped lesson, by country: 1999 92

TABLE 5.1. Average percentage of problems per eighth-grade mathematics lesson and percentage of lessons with at least one problem in which more than one solution method was presented publicly, by country: 1999 94

TABLE 5.2. Average percentage of problems per eighth-grade mathematics lesson and percentage of lessons with at least one problem in which students had a choice of solution methods, by country: 1999 95

TABLE 5.3. Average percentage of "examining methods" problems per eighth-grade mathematics lesson and percentage of lessons with at least one "examining methods" problem, by country: 1999 96

TABLE 5.4. Average percentage of problems per eighth-grade mathematics lesson that were summarized, by country: 1999 97

FIGURE 5.8. Average percentage of problems per eighth-grade mathematics lesson of each problem statement type, by country: 1999 99

FIGURE 5.9. Average percentage of problems per eighth-grade mathematics lesson solved by explicitly using processes of each type, by country: 1999 101

FIGURE 5.10. Average percentage of using procedures problems per eighth-grade mathematics lesson solved by explicitly using processes of each type, by country: 1999 102

FIGURE 5.11. Average percentage of stating concepts problems per eighth-grade mathematics lesson solved by explicitly using processes of each type, by country: 1999 103

FIGURE 5.12. Average percentage of making connections problems per eighth-grade mathematics lesson solved by explicitly using processes of each type, by country: 1999 104

FIGURE 5.13. Average percentage of private work time per lesson devoted to repeating procedures and something other than repeating procedures or mix, by country: 1999 105

TABLE 5.5.	Average percentage of non-problem segments per lesson coded as mathematical information, contextual information, mathematical activity, and announcements, by country, 1999 107
FIGURE 5.14.	Average number of teacher and student words per eighth-grade mathematics lesson, by country: 1999 109
FIGURE 5.15.	Average number of teacher words to every one student word per eighth-grade mathematics lesson, by country: 1999 110
FIGURE 5.16.	Average percentage of teacher utterances of each length per eighth-grade mathematics lesson, by country: 1999 111
FIGURE 5.17.	Average percentage of student utterances of each length per eighth-grade mathematics lesson, by country: 1999 112
TABLE 5.6.	Percentage of eighth-grade mathematics lessons during which a chalkboard, projector, textbook/worksheet, special mathematics material, and real-world object were used, by country: 1999 114
FIGURE 5.18.	Percentage of eighth-grade mathematics lessons during which computational calculators were used, by country: 1999 115
TABLE 6.1.	Average scores on the TIMSS mathematics assessment, grade 8, by country: 1995 and 1999 120
FIGURE 6.1.	Australian eighth-grade mathematics lesson signature: 1999 127
FIGURE 6.2.	Czech eighth-grade mathematics lesson signature: 1999 130
FIGURE 6.3.	Hong Kong SAR eighth-grade mathematics lesson signature: 1999 ...134
FIGURE 6.4.	Japanese eighth-grade mathematics lesson signature: 1995 137
FIGURE 6.5.	Dutch eighth-grade mathematics lesson signature: 1999 140
FIGURE 6.6.	Swiss eighth-grade mathematics lesson signature: 1999 143
FIGURE 6.7.	U.S. eighth-grade mathematics lesson signature: 1999 146
TABLE 6.2.	Similarities and differences between eighth-grade mathematics lessons in Japan and Hong Kong SAR on selected variables: 1995 and 1999 150

CHAPTER 1
Introduction

The broad purpose of the 1998–2000 Third International Mathematics and Science Study Video Study (hereafter, TIMSS 1999 Video Study) was to investigate and describe teaching practices in eighth-grade mathematics and science in a variety of countries. It is a supplement to the TIMSS 1999 student assessment, a successor to TIMSS 1995.[1] The TIMSS 1999 Video Study expanded on the earlier 1994–1995 (hereafter 1995) TIMSS Video Study (Stigler et al. 1999) by investigating teaching in science as well as mathematics and sampling classroom lessons from more countries than the TIMSS 1995 Video Study. Although data were collected at the same time, the mathematics and science portions of the TIMSS 1999 Video Study are reported separately. The results for the mathematics portion are presented in this report. Results for the science portion will be published at a later date.

The TIMSS 1995 Video Study included only one country with a relatively high score in eighth-grade mathematics as measured by TIMSS—Japan. It was tempting for some audiences to prematurely conclude that high mathematics achievement is possible only by adopting teaching practices like those observed in Japan. The TIMSS 1999 Video Study addressed this issue by sampling eighth-grade mathematics lessons in more countries—both Asian and non-Asian countries—where students performed well relative to the United States on the TIMSS 1995 mathematics assessments.[2] Countries participating in the mathematics portion of the TIMSS 1999 Video Study were Australia, the Czech Republic, Hong Kong SAR,[3] the Netherlands, Switzerland, and the United States. Japan, which participated in the science portion of the TIMSS 1999 Video Study, did not participate in the mathematics portion. However, the Japanese mathematics lessons collected for the TIMSS 1995 Video Study were re-analyzed as part of the TIMSS 1999 Video Study and are included in many of the analyses presented in this report.

In addition to the broad purpose of describing teaching in seven countries, including a number with records of high achievement in eighth-grade mathematics, the TIMSS 1999 Video Study had the following research objectives:

- To develop objective, observational measures of classroom instruction to serve as appropriate quantitative indicators of teaching practices in each country;

---

[1] TIMSS was conducted in 1994–95 and again in 1998–99. For convenience, reference will be made to TIMSS 1995 and TIMSS 1999 throughout the remainder of the report. In other documents, TIMSS 1999 is also referred to as TIMSS-R (TIMSS-Repeat).

[2] All of the countries that participated in the study performed well on the TIMSS 1995 grade 8 mathematics assessment in comparison to the United States in 1995 and were, in most cases, among the top-performing nations. Based on the results of the TIMSS 1999 grade 8 mathematics assessment, only the Czech Republic experienced a significant change in achievement between 1995 and 1999. The average mathematics achievement of eighth-graders in the Czech Republic was lower in 1999 than in 1995, and was not measurably different from the United States (Gonzales et al. 2000).

[3] For convenience, in this report Hong Kong SAR is referred to as a country. Hong Kong is a Special Administrative Region (SAR) of the People's Republic of China.

- To compare teaching practices among countries and identify similar or different lesson features across countries; and

- To describe patterns of teaching practices within each country.

Building on the interest generated by the TIMSS 1995 Video Study, the TIMSS 1999 Video Study had a final objective regarding effective use of the information:

- To develop methods for communicating the results of the study, through written reports and video cases, for both research and professional development purposes.

The TIMSS 1999 Video Study was funded by the National Center for Education Statistics (NCES), the U.S. Department of Education's Fund for the Improvement of Education, and the National Science Foundation (NSF). It was conducted under the auspices of the International Association for the Evaluation of Educational Achievement (IEA), based in Amsterdam, the Netherlands. Support for the project also was provided by each participating country through the services of a research coordinator who guided the sampling and recruiting of participating teachers. In addition, Australia and Switzerland contributed direct financial support for data collection and processing of their respective sample of lessons.

The current report focuses on the findings of the mathematics portion of the TIMSS 1999 Video Study with brief descriptions of the methods used (see appendix A). A detailed description of the methods of the mathematics portion of the TIMSS 1999 Video Study will be presented in an accompanying technical report released separately (Jacobs et al. forthcoming). A report on analyses of the science lessons also will be released separately, along with a supplementary technical report focusing on the science component of the study. A report focusing on comparisons of eighth-grade mathematics teaching in the United States based on the data collected for both the 1995 and 1999 Video Studies is also planned. The results in this report are presented from an international perspective. Individual countries may choose to issue country-specific reports at a later date.

Studying Classroom Teaching Across Countries

Why Study Teaching?

The reason for conducting a study of teaching is quite straightforward: to better understand, and ultimately improve, students' learning, one must examine what happens in the classroom. The classroom is the place intentionally designed to facilitate students' learning. Although relationships between classroom teaching and learning are complicated, it is well documented that teaching makes a difference in students' learning (Brophy and Good 1986; Hiebert 1999; National Research Council 1999). Research on teaching can stimulate discussions of ways to improve classroom learning opportunities for students.

Observing that teaching influences students' learning is not the same as claiming that teaching is the sole cause of students' learning. Many factors, both inside and outside of school, can affect students' levels of achievement (National Research Council 1999; Floden 2001; Wittrock 1986). In particular, eighth-graders' achievement in mathematics is the culmination of many past and current factors. For these reasons, no direct inferences can or should be made to link descriptions of teaching in the TIMSS 1999 Video Study with students' levels of achievement as documented in TIMSS 1999 (Martin et al. 2000; Mullis et al. 2000). Moreover, in most of the participating countries the videotaped classrooms are not the same ones in which students took the achievement tests.

Why Study Teaching in Different Countries?

If direct connections between classroom teaching and learning are difficult to draw, and if it is premature to conclude that the instructional practices used in high-achieving countries are responsible for students' learning, why study teaching in different countries and why bother to select countries with high levels of achievement? This is a key question because some readers might expect to find in this report descriptions of "correct" or at least "exemplary" teaching. That is, why can't researchers just identify the high-achieving countries, videotape their classrooms, and then assume that the teaching they see reflects best practice? As noted above, it's not that simple. This being the case, why bother to study these instructional practices? There are at least four reasons.

Reveal one's own practices more clearly

When everyday routines and practices are so culturally common that most people do things in the same way, they can become invisible (Geertz 1984). To the extent they are noticed, everyday practices can appear as the natural way to do things rather than choices that can be re-examined. A powerful way to notice the practices of one's own culture is by observing others, which is a common outcome of cross-cultural, comparative research (Ember and Ember 1998; Spindler 1978; Whiting 1954). Just as when people travel abroad and encounter strikingly different social roles and norms, comparative studies can reveal what is taken for granted in one's own culture.

So it also seems with teaching. Studying teaching practices different from one's own can reveal taken-for-granted and hidden aspects of teaching (Stigler and Hiebert 1999; Stigler, Gallimore, and Hiebert 2000). Because seeing one's own practices is a first step toward re-examining them (Carver and Scheier 1981; Tharp and Gallimore 1989), and ultimately improving them, this is a non-trivial benefit of cross-cultural and comparative studies of teaching.

Discover new alternatives

Looking at other cultures might not only help to see oneself more clearly, it might also suggest alternative practices. Although variation exists within cultures, truly distinctive teaching practices are the exception, by definition. Based on different beliefs and different expectations, teachers in other cultures might have developed entirely different teaching practices. This is, in fact, what was seen in the TIMSS 1995 Video Study (Stigler et al. 1999). Teaching in Japan looked different from teaching in Germany or the United States. Japanese teachers frequently posed mathematics problems that were new for their students and then asked them to develop a solution method on their own. After allowing time to work on the problem, Japanese teachers engaged students in presenting and discussing alternative solution methods and then teachers summarized the mathematical points of the lesson. Especially revealing was the way in which these features were combined into a pattern or system that characterized a distinctive method of teaching observed in eighth-grade Japanese mathematics lessons. These features of practice offer an alternative to those seen in the United States and Germany.

Stimulate discussion about choices within each country

Alternative practices discovered in other countries might not transpose readily across cultures. They might be based on cultural conditions that do not exist in other countries. But, seeing oneself more clearly by comparing practices across cultures, and seeing alternative practices, can underscore the idea that classroom practices are the result of choices; they are not inevitable. Choices that have been made in the past can be re-examined in a new light.

Statistical findings about between-country differences in teaching practices, along with videotapes illustrating the nature of these practices, can promote public discussion about classroom teaching. Why are particular teaching practices so common, should these methods be retained, what other choices can be made, and what conditions might support the move toward different teaching practices? These questions can be addressed with new eyes and with new information.

Deepen educators' understanding of teaching

Although research on teaching, and specifically research on mathematics teaching, has a long history (e.g., see the assortment of "Handbooks" that address issues of teaching such as Bishop et al. 1996; English 2002; Grouws 1992; Richardson 2001; Wittrock 1986), it still is difficult to form research-based hypotheses about the specific features of teaching that most influence students' opportunities to learn. This is likely due, in part, to the complex interactions among features and the different kinds of learning that different configurations of features support. Consider, for example, the number of mathematics problems that are solved during a single lesson. Is students' learning facilitated more by solving lots of problems or by solving few problems? It seems to depend on the nature and rigor of the problem content, the learning goals established, the teaching practices used, and how the problems are solved (Hiebert and Wearne 1993; Leinhardt 1986).

Cross-cultural studies of teaching provide information about different systems of teaching and different ways in which the basic ingredients of teaching can be configured (Stigler et al. 2000). Comparative findings can help researchers construct more informed hypotheses about the ways different instructional practices might influence learning. These hypotheses can then form the basis of future research that specifically seeks to determine what matters.

Why Study Teaching Using Video?

Traditionally, attempts to measure classroom teaching on a large scale have used teacher questionnaires. Questionnaires are economical and simple to administer to large numbers of respondents and usually can be transformed easily into data files that are ready for statistical analysis. However, using questionnaires to study classroom practices is problematic because it can be difficult for teachers to remember classroom events and interactions that happen quickly, perhaps even outside of their conscious awareness. Moreover, different questions can mean different things to different teachers (Stigler et al. 1999).

Direct observation of classrooms overcomes some of the limitations of questionnaires but important limitations remain. Significant training problems arise when used across large samples, especially across cultures. A great deal of effort is required to assure that different observers are recording behavior in comparable ways. In addition, and like questionnaires, the features of teaching being investigated must be decided ahead of time. Although new categories might occur to observers during the study, the earlier lessons cannot be re-observed.

Video offers a promising alternative for studying teaching (Stigler et al. 2000). Although videotaping classroom lessons brings its own challenges, the method has significant advantages over other means of recording data for investigating teaching.

Video enables the study of complex processes

Classrooms are complex environments, and teaching is a complex process. Humans can attend to a limited amount of information at any one time. It is impossible to detect, in real time, all of the important classroom events and interactions. By using video it is possible to capture the simultaneous presentation of curriculum content and execution of teaching practices. Video enables investigators to parse data analysis into more manageable portions. Observers can code video in multiple passes, coding different dimensions of teaching in each pass. And each pass can be slowed down—by viewing the same event many times. This allows coders to describe what is happening in greater detail than if they were conducting live observations, and thus permits a greater variety of analyses.

Video increases inter-rater reliability, decreases training difficulties

Video also enables solutions to problems of inter-rater reliability that are difficult to resolve in the context of live observations, especially with cross-cultural studies and classrooms that are thousands of miles apart. Researchers from different geographic locations and different cultural and linguistic backgrounds can work together, in the same location, to develop codes and establish their reliability using a common set of video data. Inter-rater disagreements can be resolved based on re-viewing the video, turning such disagreements into valuable training opportunities. And, the same segments of video can be used for training all observers, increasing the likelihood that coders will apply definitions in comparable ways.

Video enables coding from multiple perspectives

Teaching is so complex, especially when spread across seven countries, that no one person has the knowledge and skills to analyze it fully. Video allows researchers with different areas of expertise and points of view to examine the same lessons. The eighth-grade mathematics video lessons collected for this study were analyzed by native speakers from each country who were familiar with schooling in each country, mathematics educators who have studied learning and teaching in the middle grades, mathematicians who were familiar with educational issues, and specialists in language translations and linguistic analyses.

Video stores data in a form that allows new analyses at a later time

Most survey data lose their value over time. Researchers decide what questions to ask and how to code responses based on theories that are prevalent at the time. When new questions arise, conventional data, such as questionnaires, provide only limited opportunities to conduct follow-up investigations. In contrast, video data provide a relatively cost-effective way to conduct later investigations that focus on questions entirely different than those addressed when the video recordings were collected. Video data can be re-coded in many different ways, for many different purposes, and analyzed as theories and questions change over time, giving them a longer shelf life than data recorded on paper. Re-coding the Japanese sample of videotapes collected in the TIMSS 1995 Video Study with the new TIMSS 1999 coding scheme illustrates this benefit.

Video facilitates integration of qualitative and quantitative information

Video makes it possible to merge qualitative and quantitative analyses in a way not possible with other kinds of data. This often occurs through a "cycle of analysis" that continually links qualitative descriptions with quantitative coding and analysis (Jacobs, Kawanaka, and Stigler 1999). In this study, the process often began with a qualitative analysis of a few lessons. In-depth discussions produced hypotheses, informed by previous research, about the comparative nature of teaching across countries or about the relations among parts of lessons within countries. Hypotheses were then tested quantitatively by coding a larger sample of lessons. Results were examined and hypotheses were refined or abandoned, and new questions were asked. After multiple codes were applied and results analyzed quantitatively, qualitative descriptions were used to recompose the findings into meaningful constructs. In this way, the cycle moved from observing to generating to coding to evaluating and then full circle back to observing.

Video facilitates communication of the results

It is also possible, with video, to report research results using concrete, "real" examples. The video clips that accompany this report provide a richer sense of what the codes mean and a concrete basis for interpreting the quantitative findings. The clips provide observable definitions of many of the codes used to analyze the lessons (see CD-ROM that accompanies this report). Such video-enhanced definitions can, over time, provide educators with a set of shared referents for commonly used descriptors, such as "problem solving." This could yield a shared language of classroom practice, an essential tool in building a widely shared professional knowledge base for teaching. In the long run, a shared set of referents even could lead to the development of more efficient and valid research instruments, including questionnaire-based indicators of instructional quality.

In addition to the video clips accompanying this report, sample lessons for public release were collected as part of the TIMSS 1999 Video Study. These lessons will be widely available on CD-ROMs and other media, and will include video-linked commentary by the teacher and by educators within the respective country. The goal is to provide teachers around the world with samples of the kind of lessons that were analyzed as part of TIMSS 1999 Video Study and to stimulate local and international discussions of mathematics teaching.

What Are the Challenges of Studying Teaching Using Video?

Video data bring their own set of challenges. A brief review of how these challenges were addressed in this study is provided below. For a more complete discussion, see the TIMSS 1999 Video Study technical report (Jacobs et al. forthcoming).

Standardization of camera procedures

Deciding exactly what to film during a lesson is a nontrivial issue. To study classroom teaching in a consistent way across classrooms, it is important to develop standardized procedures for using the camera and then to carefully train videographers to follow these procedures. In this study, the camera followed what an attentive student would be looking at during times of public discussion, usually the teacher, and then followed the teacher and sampled students' activities during private work time. A second camera was stationary and maintained a wide-angle shot of the students.

Observer effects

What effect does the camera have on what happens in the classroom? Will students and teachers behave as usual with the camera present, or will the camera capture a view that is biased in some way? To minimize camera effects, teachers were asked to teach as usual and to carry out the lesson they would have taught had the videographer not been present. After filming, the teachers provided written responses to questions that permitted an assessment of the lesson's typicality. Teachers were asked, for example, to describe the lessons they taught to the same class the day before and the day after the filmed lesson, and they were asked to comment on any unusual features in the filmed lesson. Teachers knew ahead of time that they would be filmed, so they probably tried to do an especially good job and might have done some extra preparation. But teachers are likely to be constrained by what students expect and by their own repertoire of teaching practices. Videotaped lessons probably are best interpreted as a slightly idealized version of what the teacher typically does in the classroom.

Sampling and validity

Due to the expense of filming lessons around the world, there is a limit to how many lessons can be included. Sampling becomes an important issue. How many teachers should be selected, and how many lessons per teacher should be filmed? The answers depend on the goal of the study and the level of analysis to be used. If researchers need a valid and reliable picture of individual teachers, then teachers must be taped multiple times. Teachers can vary from day to day in the kind of lesson they teach, as well as in the success with which they implement the lesson. If, on the other hand, researchers want a school-level picture, or a national-level picture, then each teacher can be taped fewer times. What is essential in these cases is that a sufficient number of different teachers are included. Of course, if teachers are taped only one or a few times, researchers and interpreters must resist the temptation to view these data as reliable descriptions of individual teachers.

The goal of the TIMSS 1999 Video Study was to provide national-level pictures of teaching. Consequently, it was important to invest the finite resources in maximizing the number of teachers even though this meant videotaping each teacher only once, teaching a single classroom lesson.

Taping only one lesson per teacher shapes the kinds of conclusions that can be drawn about instruction from this study. Teaching involves more than constructing and implementing lessons. It also involves weaving together multiple lessons into units that can stretch out over days and weeks. Inferences about the full range of teaching practices and dynamics that might appear in a unit cannot necessarily be made, even at the aggregate level, based on examining a single lesson per teacher. Consequently, the interpretive frame of the TIMSS 1999 Video Study is properly restricted to national-level descriptions and comparisons of individual lessons.

Another sampling issue concerns the way in which content is sampled. Because different countries teach somewhat different topics in eighth-grade mathematics and teach them at different times of the year, the best strategy for this study was to randomly select lessons across the school year.[4]

---

[4] The sample of Japanese lessons collected for the TIMSS 1995 Video Study and re-analyzed as part of this study did not include lessons drawn from across the full school year; most were collected over a four-month period (Stigler et al. 1999). Analysis of the Japanese mathematics data (explicated in later sections of this report) does not reveal any systematic problems with the representativeness of the findings. Therefore, while the sampling of the lessons was less than ideal, there is no evidence that the Japanese eighth-grade mathematics data are not representative of teaching at that time.

Coding reliability

As with all observational studies, the importance of clearly defining and applying codes to the data, and then making sure that the coders are categorizing the data as consistently and accurately as possible, is paramount. Some behaviors or activities can be easier to define and identify in videotape, such as whether a teacher uses an overhead projector. Either the teacher does or does not use one. On the other hand, there are numerous other aspects of teaching mathematics that can be more difficult to define and to identify consistently, especially across sets of videotapes from different countries that are being processed by teams of coders who have different experiences and expertise, speak different languages, and may bring to the process different cultural and social expectations. Because of this, those applying codes to the videotape data went through a rigorous training procedure prior to applying the codes to the main data sets, to achieve high rates of agreement between the coders.

There are various methods for establishing the reliability of coding procedures, depending on the type of data being coded and the categories being applied. In this study, there were two aspects of coding that were evaluated throughout the coding process. First, for a number of codes, it was important to assess the degree to which coders applied a code at or very near the same point of time in the lesson. That is, because in some instances it was important not only to know that an activity occurred during the lesson but also how long that activity took place, the success of the coding process was evaluated by measuring the accuracy of coding the in- and out-points in the timing of an activity on the videotape. Second, and as is customary practice in observational studies, it was important to assess whether different coders applied the same codes to the same behaviors or activities occurring during the lesson. All three marks (i.e., the in-point, out-point, and category) were evaluated and included in the measures of reliability.

A common measure, known as percentage agreement, was used to measure both inter-rater reliability and code reliability within and across the countries. Percentage agreement is calculated by dividing the number of agreements by the number of agreements plus disagreements. The percentage agreement for each code at the beginning and midpoint of the coding process is included in appendix A in the back of the report and is also documented in greater detail in the forthcoming technical report (Jacobs et al.). For all codes, the minimum acceptable reliability score, averaging across coders, was 85 percent. Moreover, the minimum acceptable reliability score for an individual coder or coder pair was 80 percent. Reliability among coders was measured constantly throughout the months-long coding process to ensure that coders continually met the minimum acceptable standard. Because so few lessons were coded at the very beginning of the process, initial reliability among coders and across countries was measured against a set of master lessons, which were coded and agreed to by the entire mathematics code development team. As more and more lessons were coded, it was then easier to measure reliability between pairs of coders. If a coder did not meet the minimum reliability standard, more training was provided until an acceptable level was achieved. If, after numerous attempts, reliability measures fell below the minimum acceptable standard (as described above), the code was dropped from the study. The procedures used to measure reliability are described in Bakeman and Gottman (1997).

The unreasonable power of the anecdote

One of the biggest boons of video data also is their bane. Video images are vivid and powerful tools for representing and communicating information. But video images can be too powerful. One video image, although memorable, can be misleading and unrepresentative of reality. This becomes a problem because humans easily can be misled by anecdotes, even in the face of contradictory and far more valid information (Nisbett and Ross 1980). In fact, methods of research design and inferential statistics were developed specifically to protect people from being misled by anecdotes and personal experiences (Fisher 1951).

Fortunately video surveys provide a way to resolve the tension between anecdotes (visual images) and statistics (Stigler et al. 2000). Discoveries made through qualitative analysis of a few videos can be validated by statistical analysis of the whole set. For example, while watching a video the researchers might notice an interesting technique used by an Australian teacher. If they had only one video, they would not know what to make of this observation: do Australian teachers use the technique on a regular basis, more than teachers in other countries, or did they just happen to notice one powerful example in the Australian data? Because the TIMSS 1999 Video Study collected a large sample of lessons, researchers could turn their observations into hypotheses that could be validated against the database.

In a complementary process, the research team might, after coding and analyzing the quantitative video data, discover a statistical relationship in the data. By returning to the actual videos, they could find concrete images to attach to their discovery, giving a means for further analysis and exploration, as well as a set of powerful images that can be used to communicate the statistical discovery. Through this process, the statistic can be brought to life.

Building on the TIMSS 1995 Video Study

Film was first used in cultural and educational studies in the 1930s. Video recording began to be used as soon as technological development made it practical to do so. However, these applications were ethnographic in nature and none employed nationally representative samples and multiple cultures (de Brigard 1995; National Research Council 2001b; Spindler and Spindler 1992). The TIMSS 1995 Video Study was the first study to use video technology to investigate classroom teaching on a country-wide basis and compare teaching across countries (Stigler et al. 1999). Great strides were made in the TIMSS 1995 Video Study for dealing with the considerable logistical and methodological challenges of conducting a large-scale, international video survey. These include procedures for videotaping classrooms, processing and storing videos for easy access, and the development of software that links video to transcripts and permits coding to be done as researchers view the lesson videos (Knoll and Stigler 1999).

The TIMSS 1999 Video Study built on the earlier video study in another way. A central hypothesis emerging from the TIMSS 1995 Video Study is that there are distinct patterns of mathematics teaching in different countries (Stigler and Hiebert 1999). This result had not been anticipated, and so it had not been addressed in the design of the original study. In contrast, the TIMSS 1999 Video Study research team began by soliciting tentative descriptions of typical lessons from experts in each country and used these descriptions to frame the development of a coding system that would capture features of eighth-grade mathematics teaching considered essential from each country's perspective. These typical lessons, or country models of teaching, were continually revisited to ensure that each country's perspective was considered as individual codes were constructed. A description of country model development and the models themselves are presented in appendix E.

Codes developed in the earlier video study also provided a base on which the TIMSS 1999 Video Study could build. Although expanding the sample to seven countries made it impossible to retain many of the exact codes from the TIMSS 1995 Video Study, it was possible to preserve key ideas such as examining the organization of lessons, the nature of the mathematics presented, and the way in which mathematics was worked on during the lesson.

The TIMSS 1995 Video Study provided a starting point for this study, both in methods and substance. To extend the 1995 study and address the question of whether eighth-grade teachers in higher achieving countries teach mathematics in similar ways, the TIMSS 1999 Video Study included more high-achieving countries, based on results of TIMSS 1995 assessments. Table 1.1 lists the countries that participated in the TIMSS 1999 Video Study along with their scores on TIMSS 1995 and TIMSS 1999 mathematics assessments. On the TIMSS 1995 mathematics assessment, eighth-graders as a group in Japan and Hong Kong SAR were among the highest achieving students, and their results were not found to be significantly different from one another (Beaton et al. 1996). Students in the Czech Republic scored on average significantly below their peers in Japan, but no differences were detected from the average score in Hong Kong SAR. Average scores in Switzerland and the Netherlands were also not found to be different from one another. The mathematics average for Australia was not detectably different from the average score in the Netherlands. Eighth-grade students in the United States scored, on average, significantly lower than their peers in the other six countries in 1995.

The TIMSS 1999 mathematics assessment was administered after the TIMSS 1999 Video Study was underway and played no role in the selection of countries for the Video Study. However, the TIMSS 1999 mathematics results indicated that eighth-graders in countries participating in the TIMSS 1999 Video Study continued to score significantly higher than their peers in the United States, except for students in the Czech Republic whose scores were not significantly different than students in the United States as a result of a significant decline in the average mathematics score between 1995 and 1999 in the Czech Republic. Switzerland did not participate in the TIMSS 1999 assessment.

TABLE 1.1. TIMSS 1999 Video Study participating countries and their average score on TIMSS 1995 and TIMSS 1999 mathematics assessments

Country	TIMSS 1995 mathematics score[1] Average	TIMSS 1995 mathematics score[1] Standard error	TIMSS 1999 mathematics score[2] Average	TIMSS 1999 mathematics score[2] Standard error
Australia[3] (AU)	519	3.8	525	4.8
Czech Republic (CZ)	546	4.5	520	4.2
Hong Kong SAR (HK)	569	6.1	582	4.3
Japan (JP)	581	1.6	579	1.7
Netherlands[3] (NL)	529	6.1	540	7.1
Switzerland (SW)	534	2.7	—	—
United States (US)	492	4.7	502	4.0
International average[4]	—	—	487	0.7

—Not available.
[1]TIMSS 1995: AU>US; HK, JP>AU, NL, SW, US; JP>CZ; CZ, SW>AU, US; NL>US.
[2]TIMSS 1999: AU, NL>US; HK, JP>AU, CZ, NL, US.
[3]Nation did not meet international sampling and/or other guidelines in 1995. See Beaton et al. (1996) for details.
[4]International average: AU, CZ, HK, JP, NL, US>international average.
NOTE: Rescaled TIMSS 1995 mathematics scores are reported here (Gonzales et al. 2000). Due to rescaling of 1995 data, international average not available. Switzerland did not participate in the TIMSS 1999 assessment.
SOURCE: Gonzales, P., Calsyn, C., Jocelyn, L., Mak, K., Kastberg, D., Arafeh, S., Williams, T., and Tsen, W. (2000). *Pursuing Excellence: Comparisons of International Eighth-Grade Mathematics and Science Achievement From a U.S. Perspective, 1995 and 1999* (NCES 2001-028). U.S. Department of Education. Washington, DC: National Center for Education Statistics.

What Can Be Found in This Report?

The goal of teaching is to facilitate learning. Correspondingly, the overriding goal of this study is to describe aspects of teaching that appear to be designed to influence students' learning opportunities. The presentation of results is organized around the following three aspects of teaching that both seem to contribute to students' learning opportunities and were found in the TIMSS 1995 Video Study to distinguish among countries in terms of teaching practices: the way lessons were organized or, said another way, the way the learning environment was structured (chapter 3); the nature of the content of the lessons (chapter 4); and the instructional practices, or ways in which the content was worked on during the lessons (chapter 5). Before presenting the findings from the video lessons, however, it is useful to learn something about the participating teachers and their view of the filmed lesson (chapter 2). The report concludes by stepping back and considering what conclusions can be drawn from the patterns evident across the individual features of teaching (chapter 6).

Chapter 2 focuses on the context of the lessons as reported by the participating teachers' responses to the written questionnaire. The findings address questions such as:

- How prepared were participating teachers to teach eighth-grade mathematics?

- How typical was the filmed lesson?

- What were the goals for the lesson?

- How aware were the teachers of current ideas for teaching mathematics and did they perceive the filmed lesson to be consistent with these ideas?

Chapter 3 includes results on the structure of the lesson and the classroom. These data set the stage, in many ways, for the details of the lesson activities discussed in later chapters. Findings in chapter 3 address questions such as:

- How long did students spend studying mathematics?

- How were the lessons divided among activities that focused on review, introducing new material, and practicing new material?

- How was the classroom organized in terms of whole-class discussion and individual student work?

- What role did homework play in the lesson?

Chapter 4 examines the content of the lesson. Questions addressed in chapter 4 include:

- What mathematical topics were covered in the lessons?

- How complex was the mathematics?

- What kinds of mathematical reasoning were encouraged by the problems presented?

- How was the content related across the lesson?

Chapter 5 considers the way in which mathematics was worked on by the teacher and students during the lesson. The questions addressed by the findings in this chapter include:

- In what contexts were mathematical problems presented (e.g., were they embedded in real-life situations)?

- Could students choose the methods they wished to use to solve mathematical problems, and were multiple solution methods presented?

- What was the relationship between the kinds of mathematical problems presented and the way the teacher and students worked through them?

- What kind of mathematical work were students expected to do when they worked on their own?

Chapter 6 concludes the report by pulling together the individual features of teaching presented earlier and answering the major questions addressed by the study. The concept of a "lesson signature" is introduced to capture the way in which the basic ingredients of lessons were put together across real time in each country to create patterns or systems of teaching. The questions that are addressed in this final chapter include:

- Are there similarities in eighth-grade mathematics teaching across the seven countries?

- What are the distinctive characteristics of eighth-grade mathematics teaching in each country?

For all analyses presented in this report, differences between averages or percentages that are statistically significant are discussed using comparative terms such as "higher" and "lower." Generally, differences that are not found to be statistically significant are not discussed, unless warranted. To determine whether differences reported are statistically significant, ANOVAs and two-tailed t-tests, at the .05 level, were used. Bonferroni adjustments were made when more than two groups were compared simultaneously (e.g., a comparison among all seven countries). The analyses were conducted using data weighted with survey weights, which were calculated specifically for the classrooms in the TIMSS 1999 Video Study. The weights were developed for each country, so that estimates are unbiased estimates of national means and distributions. The weight for each classroom reflects the overall probability of selection for that classroom, with appropriate adjustments for non-response (see the technical report, Jacobs et al. forthcoming, for a more detailed description of weighting procedures). In some cases, large apparent differences in data are not significant due to large standard errors, small sample sizes, or both. Standard errors for all estimates displayed in the figures and tables in the report are included in appendix C.

To assist readers in interpreting data presented in tables and figures, results of the statistical tests are listed below each table and figure in which data are compared. Results are indicated by the use of the greater than (>) symbol, e.g., AU>CZ for Australia's average is greater than the Czech Republic's average. Only those comparisons that were determined to be significant are listed.

Accompanying this report is a CD-ROM on which short video clips that illustrate many of the codes used to analyze the lesson videos are presented. The video clips are taken from lessons filmed specifically for the purpose of public display. These lessons were not included in the samples collected in each country and analyzed for this report. Permission to display the video clips was granted by all participants and/or their legal guardians in these public release videotape lessons. The CD-ROM is entitled "Teaching Mathematics in Seven Countries Video Clip Examples." In chapters 3, 4, and 5 of the published version of this report, a camera icon (📹) and note is provided that indicates the number of the video clip on the CD-ROM relevant to the discussion. For the CD-ROM version of this report, a hyperlink to the relevant example is provided.

Also released simultaneously with this report is a brochure that discusses study highlights (*Highlights From the TIMSS 1999 Video Study of Eighth-Grade Mathematics Teaching*, NCES 2003-011) and four full-length lesson videos from each of the 7 participating countries. These 28 public release videos are presented as a set of CD-ROMs and include, in addition to lesson videos, accompanying materials including a transcript in English and the native language, and commentaries by teachers, researchers, and national research coordinators in English and the native language. These public release videos and materials are intended to augment the research findings, support teacher professional development programs, and encourage wide public discussion of teaching and how to improve it.

All of these products can be accessed or ordered by going to the NCES web site (*http://nces.ed.gov/timss*).

A Lens Through Which to View This Report

The mathematics portion of the TIMSS 1999 Video Study moved beyond the TIMSS 1995 Video Study in several ways, including a larger sample of countries, the development of a new coding scheme to analyze classroom lessons, and the recruitment of additional specialists to assist with the analysis. It is hoped that the most significant extension, however, will come as readers digest the contents of this report and engage in deeper and more nuanced international discussions of mathematics teaching.

The descriptions of classroom lessons presented here reveal a complex variety of features and patterns of teaching. They are similar to, and different from, each other in interesting and sometimes subtle ways. Whereas the TIMSS 1995 Video Study revealed the striking case of Japan and seemed to lead the casual reader (or viewer) into presuming that a "Japanese method" of teaching is necessary for high achievement, no similarly easy interpretation of the TIMSS 1999 Video Study results is possible. It will be seen that countries with high achievement teach in a variety of ways. Interpretation of these results requires a thoughtful and analytic approach. This study did not attempt to examine best teaching practices and its results do not identify which practices yield high achievement. But it is reasonable to search these results for different choices that teachers in different countries have made in order to more clearly see teaching in one's own country. If these more complex results translate into increasingly rich and productive examinations and discussions of teaching, both within and across countries, then it is possible that progress will be made toward understanding and improving teaching.

CHAPTER 2
Context of the Lessons

This chapter presents the results of teacher responses to questionnaire items designed to provide background information on the videotaped teachers and to help assess the typicality of the videotaped lesson. Questionnaire data were obtained from teachers in 100 percent of the eighth-grade mathematics lessons videotaped in Australia, the Czech Republic, Hong Kong SAR, and the United States, 96 percent of Dutch lessons, and 99 percent of Swiss lessons. Questionnaire data were collected in Japanese mathematics lessons as part of the TIMSS 1995 Video Study using a different version of the teacher questionnaire. Results from the Japanese teacher questionnaire data are presented in Stigler et al. (1999) and are not included here.[1]

There are many factors that define the context of an eighth-grade mathematics lesson. These include, among other things, characteristics of the teachers, their expectations for mathematics teaching and learning, and where the lesson fits in the curricular sequence. To collect information on these factors, questionnaire items addressed the following topics:

- Teachers' background experiences and workload;
- Teachers' learning goals for the videotaped lessons;
- Teachers' current ideas about teaching and learning mathematics; and
- Teachers' perceptions of the typicality of the videotaped lesson.

The Teachers

Teachers' Background Experiences and Workload

Mathematics teachers bring a variety of educational and professional experiences to the classes they teach. These experiences can influence their planning and implementation of a lesson (Fennema and Franke 1992; National Research Council 2001a). To better understand the eighth-grade mathematics lessons of teachers who participated in the video study, data were collected on teachers' educational preparation, professional background, and current teaching responsibilities. When interpreting the results, the reader should keep in mind that some results could be influenced by national requirements and/or support, which could vary by country.

---

[1] More information on teacher response rates, as well as the development of the questionnaires and how they were coded, can be found in appendix A and in the forthcoming technical report (Jacobs et al. forthcoming). The questionnaires are available online at *http://www.lessonlab.com*.

Educational preparation

Teachers were asked about their training in and preparation for teaching mathematics. When applicable, teachers provided information about their major field of study in both their undergraduate and graduate studies. Teachers were free to define "major field" and to mention as many fields of study that applied. Because a teacher could have listed more than one field, responses for college and graduate studies were coded into as many categories as needed. Therefore, the percentages presented in table 2.1 could add to more than 100 percent within a country and are based on teachers who identified one or more major fields of study.[2] As table 2.1 indicates, 96 percent and 90 percent of the eighth-grade mathematics lessons in the Czech Republic and the Netherlands, respectively, were taught by teachers who reported a major field of study in mathematics or mathematics education either at the undergraduate or graduate level, or both. These represent a greater percentage of lessons than in the other countries where data are available (ranging from 41 percent in Hong Kong SAR to 64 percent in Australia) including the United States at 57 percent. Compared to all the other countries except Australia, more U.S. eighth-grade mathematics lessons (50 percent) were taught by teachers who reported a major field of study in education. Seventeen percent to 44 percent of lessons were taught by teachers who reported that their major field of study was in science or science education.

In some post-secondary institutions, students can obtain minors in various fields of study. The teachers who participated in the study were therefore asked to indicate whether they had a minor in a field in addition to a major field of study, either at the undergraduate or graduate level. When both major and minor fields of study were considered, between 83 percent and 99 percent of lessons in all the countries except Switzerland were taught by teachers who identified mathematics or mathematics education as their major or minor field of study (data not shown in figure). Fifty-eight percent of Swiss lessons were taught by teachers who identified mathematics or mathematics education as a major or minor field of study in their undergraduate or graduate studies. Across the countries, 32 percent to 51 percent of lessons were taught by teachers who identified science or science education as a major or minor field of study, and 7 percent to 55 percent of lessons were taught by teachers who identified education as either a major or minor field of study.

---

[2]The percentage of lessons taught by teachers who reported various major fields of study may be affected by the limited samples collected for this study and may differ from national statistics available from other studies. For example, data from the Schools and Staffing Survey (SASS) in the United States indicate that 49 percent of mathematics courses were taught by eighth-grade public and private school teachers with a major in mathematics or mathematics education at the undergraduate or graduate level (Schools and Staffing Survey, 1999–2000 "Public Teacher Survey," "Public Charter Teacher Survey," and "Private Teacher Survey," unpublished tabulations).

TABLE 2.1. Percentage of eighth-grade mathematics lessons taught by teachers who identified one or more major fields of undergraduate and graduate study, by country: 1999

Major field	AU	CZ	HK	NL	SW	US
			Percent			
Mathematics[2,4]	64	96	41	90	61	57
Science[3,5]	28	41	33	44	35	17
Education[6]	25	18	9	13	11	50
Other[7]	30	32	35	23	19	27

[1] AU=Australia; CZ=Czech Republic; HK=Hong Kong SAR; NL=Netherlands; SW=Switzerland; and US=United States.
[2] Mathematics includes teachers' responses indicating a major field of study in either mathematics or mathematics education.
[3] Science includes teachers' responses indicating a major field of study in science, science education, or any of the various fields of science (e.g., physics, chemistry, biology).
[4] Mathematics: CZ, NL>AU, HK, SW, US.
[5] Science: NL>US.
[6] Education: US>CZ, HK, NL, SW.
[7] Other: No differences detected.

NOTE: Percentages may not sum to 100 because teachers could identify more than one major field of study. Percentages are based on responses from teachers who identified at least one major field of study.
SOURCE: U.S. Department of Education, National Center for Education Statistics, Third International Mathematics and Science Study (TIMSS), Video Study, 1999.

The degrees teachers earn might or might not be indicative of having completed requirements for certification to teach eighth-grade mathematics in each country. To further clarify their preparation for teaching eighth-grade mathematics, teachers were asked to list what subject areas and corresponding grade levels they were certified to teach. As with major areas of study, teachers could identify more than one subject area in which they were certified and their responses were coded into as many categories as were appropriate. Each subject area a teacher mentioned also was coded for the corresponding grade level for which he or she was certified. If a subject area was mentioned, the response was divided into two mutually exclusive groups: (1) teacher's certification in this subject area included grade 8 or (2) teacher's certification in this subject area was not identified for grade level or did not include grade 8.

Table 2.2 shows the results of coding teachers' responses. At least 97 percent of lessons in any country were taught by teachers who identified one or more particular subject areas in which they were certified to teach. But not all lessons were taught by teachers who reported certification to teach eighth-grade mathematics, with percentages ranging from 48 (Switzerland) to 91 (the Netherlands). Fewer Swiss lessons were taught by teachers who reported being certified to teach eighth-grade mathematics compared to the Czech Republic, Hong Kong SAR, the Netherlands, and the United States.

Teaching Mathematics in Seven Countries
Results From the TIMSS 1999 Video Study

TABLE 2.2. Percentage of eighth-grade mathematics lessons taught by teachers certified in various subject areas, by grade level of certification and country: 1999

Subject area of certification	AU	CZ	HK	NL	SW	US
One or more subject areas identified[2]	100	99	100	97	99	97
Mathematics						
Grade 8[3]	66	85	82	91	48	79
Other/unspecified grade[4]	24	14	11	7	‡	‡
Science						
Grade 8[5]	30	38	36	39	31	20
Other/unspecified grade[6]	19	7	13	6	4	10
Education						
Grade 8[7]	‡	15	‡	‡	43	22
Other/unspecified grade[8]	10	‡	‡	‡	‡	8
Other						
Grade 8[9]	20	33	35	19	34	18
Other/unspecified grade[10]	20	4	13	‡	‡	7

‡Reporting standards not met. Too few cases to be reported.
[1]AU=Australia; CZ=Czech Republic; HK=Hong Kong SAR; NL=Netherlands; SW=Switzerland; and US=United States.
[2]One or more subject areas identified: No differences detected.
[3]Mathematics–Grade 8: CZ, HK, NL, US>SW; NL>AU.
[4]Mathematics–Other/unspecified grade: AU>NL.
[5]Science–Grade 8: No differences detected.
[6]Science–Other/unspecified grade: AU>SW.
[7]Education–Grade 8: SW>CZ, US.
[8]Education–Other/unspecified grade: No differences detected.
[9]Other–Grade 8: No differences detected.
[10]Other–Other/unspecified grade: AU>CZ, US.
NOTE: Percentages do not sum to 100 because teachers could identify more than one subject area.
SOURCE: U.S. Department of Education, National Center for Education Statistics, Third International Mathematics and Science Study (TIMSS), Video Study, 1999.

Years of teaching experience

In addition to formal education and certification, teachers bring a variety of professional experiences to their classrooms, including the number of years they have been teaching. Teachers were asked to identify how many years they had been teaching, in general, and also how many years they had been teaching mathematics. On average, eighth-grade mathematics lessons in Australia, the Czech Republic, and Switzerland were taught by teachers who reported teaching at least 17 years (table 2.3) with nearly similar average number of years specifically teaching mathematics (16, 21, and 18 years respectively). Comparatively, eighth-grade mathematics lessons in Hong Kong SAR and in the Netherlands were taught by teachers who reported fewer years teaching (10 and 13 years respectively) and specifically teaching mathematics (10 and 11 years respectively) on average than their counterparts in Australia, the Czech Republic and Switzerland. Teachers of eighth-grade mathematics lessons in the United States reported an average of 14 years teaching which is significantly less than their Czech counterparts but not measurably different from their colleagues in the other countries.

Chapter 2 | 19
Context of the Lessons

TABLE 2.3. Mean, median, and range of number of years that teachers reported teaching in general and teaching mathematics, by country: 1999

Teaching experience	Country[1]						
	AU	CZ	HK	NL	SW	US
Years teaching						
Mean[2]	17	21	10	13	19	14
Median	16	21	8	12	20	14
Range	1–38	2–41	1–34	1–33	0–40	1–40
Years teaching mathematics						
Mean[3]	16	21	10	11	18	12
Median	15	21	7	11	20	10
Range	1–38	2–41	1–34	1–32	0–39	1–40

[1]AU=Australia; CZ=Czech Republic; HK=Hong Kong SAR; NL=Netherlands; SW=Switzerland; and US=United States.
[2]Years teaching: AU, CZ, SW>HK, NL; CZ>US.
[3]Years teaching mathematics: AU, CZ, SW>HK, NL; CZ>AU, US; SW>US.
NOTE: Mean years are calculated as the sum of the number of years reported for each lesson divided by the number of lessons within a country. For each country, median is calculated as the number of years below which 50 percent of the lessons fall. Range describes the lowest number of years and the highest number of years reported within a country.
SOURCE: U.S. Department of Education, National Center for Education Statistics, Third International Mathematics and Science Study (TIMSS), Video Study, 1999.

Time spent on different school activities

Teachers have many responsibilities, both related and unrelated to their mathematics teaching. To understand some of these demands, teachers were asked to estimate the amount of time they devoted to teaching mathematics, teaching other classes, and engaging in other school-related activities during a typical week.

Table 2.4 shows that eighth-grade mathematics lessons differed on the amount of time teachers reported allocating to teaching mathematics. Lessons in the Netherlands and the United States were taught by teachers who reported spending the largest amount of time, 18 to 20 hours a week on average, teaching mathematics. Swiss lessons were taught by teachers who reported spending more time teaching classes other than mathematics—an average of 13 hours per week—compared to mathematics lessons in the other countries. Dutch lessons were taught by teachers who reported spending more time on average doing mathematics-related work at home and less time teaching other classes compared to teachers in the Czech Republic, Hong Kong SAR, and Switzerland. Dutch lessons were taught by teachers who also reported spending less time on average doing other school-related activities compared to Hong Kong SAR and Swiss teachers.

TABLE 2.4. Average hours per week that teachers reported spending on teaching and other school-related activities, by country: 1999

Activity	AU	CZ	HK	NL	SW	US
	\multicolumn{6}{c	}{Hours per week}
All teaching and other school-related activities[2]	36	42	41	39	42	42
Teaching mathematics[3]	12	14	13	20	11	18
Teaching other classes[4]	4	8	6	3	13	4
Meeting with other teachers to work on curriculum and planning issues[5]	2	1	1	1	2	2
Mathematics-related work at school[6]	6	6	9	3	3	7
Mathematics-related work at home[7]	6	6	5	8	5	6
Other school-related activities[8]	6	8	7	4	9	5

[1]AU=Australia; CZ=Czech Republic; HK=Hong Kong SAR; NL=Netherlands; SW=Switzerland; and US=United States.
[2]All teaching and other school related activities: CZ, SW, US>AU.
[3]Teaching mathematics: CZ, HK>SW; NL, US>AU, CZ, HK, SW.
[4]Teaching other classes: CZ>AU, NL, US; HK>NL; SW>AU, CZ, HK, NL, US.
[5]Meeting with other teachers to work on curriculum and planning issues: AU, SW>HK.
[6]Mathematics-related work at school: AU, CZ, HK, US>NL, SW.
[7]Mathematics-related work at home: NL>CZ, HK, SW.
[8]Other school-related activities: HK>NL; SW>NL, US.
NOTE: Average hours per week calculated by the sum of hours for each lesson divided by all lessons within a country. Hours may not sum to totals because of rounding.
SOURCE: U.S. Department of Education, National Center for Education Statistics, Third International Mathematics and Science Study (TIMSS), Video Study, 1999.

Teachers' Learning Goals for the Videotaped Lessons

A key contextual variable that shapes the nature of teaching is the set of learning goals toward which the teacher is working (Hiebert et al. 1997). Teachers were asked to describe, in open-ended questions, the "main thing" they wanted students to learn from the videotaped lesson. Some teachers listed general topical goals, such as "learning about linear systems," whereas other teachers described their goals in more detail, such as "understanding the graphical solution to linear systems: parallel lines have no common value."

Teachers' responses were evaluated along each of three dimensions: content, process, and perspective. These dimensions were based on the coding scheme developed for the TIMSS mathematics curriculum framework (Robitaille 1995; Schmidt, McKnight, Valverde, Houang, and Wiley 1997).[3] Content goals were identified by statements describing specific mathematical concepts or topics. Process goals were defined as descriptions about how teachers wanted their students to use mathematics, such as "solve equations," "solve problems," and "apply mathematics to everyday situations." Perspective goals included those aimed at promoting students' ideas and interest in mathematics and learning, such as "to be sure of their math abilities," "to see that math is fun," and "to learn to be neat and orderly in their work."

---

[3]More details about these categories can be found in appendix A and the forthcoming technical report.

Teachers' responses were coded for each dimension. For example, the response "understand the graphical solution to linear systems: parallel lines have no common value," was coded as describing a content goal (algebra) and a process goal (making connections between representations), but was not coded as describing a perspective goal. By contrast, the response "to be sure of their math abilities" was coded as describing a perspective goal (confidence in mathematical abilities) but was not coded as a content or a process goal. Results of applying this coding scheme to teachers' reported goals for the videotaped lesson are described below.

Figure 2.1 presents the percentage of eighth-grade mathematics lessons taught by teachers who identified specific content, process, or perspective goals when asked to identify their goal for the videotaped lesson. Between 75 and 95 percent of the lessons in all countries were taught by a teacher who listed a content goal for the lesson, and between 90 and 98 percent of lessons were taught by a teacher who listed a process goal for the lesson. However, between 4 and 23 percent of lessons were taught by a teacher who identified a perspective goal for the lesson.

Within-country comparisons indicated that there were no differences found in both the Czech Republic and Hong Kong SAR between the percentages of eighth-grade mathematics lessons taught by teachers who identified content and process goals for the videotaped lesson. In Australia, the Netherlands, Switzerland, and United States, a larger percentage of lessons were taught by teachers who identified process goals than content goals. Perspective goals were least common. A smaller percentage of lessons in all countries were taught by teachers who identified perspective goals than identified either content goals or process goals.

FIGURE 2.1. Percentage of eighth-grade mathematics lessons taught by teachers who identified content, process, or perspective goals for the videotaped lesson, by country: 1999

Country	Content goal[2]	Process goal[3]	Perspective goal[4]
AU	75	90	14
CZ	95	94	4
HK	93	95	‡
NL	91	98	9
SW	81	93	23
US	81	96	11

‡Reporting standards not met. Too few cases to be reported.
[1]AU=Australia; CZ=Czech Republic; HK=Hong Kong SAR; NL=Netherlands; SW=Switzerland; and US=United States.
[2]Content goal: CZ>SW.
[3]Process goal: No differences detected.
[4]Perspective goal: SW>CZ.
SOURCE: U.S. Department of Education, National Center for Education Statistics, Third International Mathematics and Science Study (TIMSS), Video Study, 1999.

Process goals for the videotaped lesson are of special interest because these goals could range from practicing routine operations (e.g., calculations and symbol manipulation) to reasoning mathematically (e.g., logical reasoning, explaining relationships). The TIMSS 1995 Video study found significant differences among the three countries in the emphasis teachers placed on developing skills versus thinking and reasoning mathematically (Stigler et al. 1999).

At least 90 percent of eighth-grade mathematics lessons in each country were taught by teachers who identified a process goal for the videotaped lessons (figure 2.1). Table 2.5 shows that, in all countries, between 40 and 51 percent of the eighth-grade mathematics lessons were taught by teachers who identified process goals related to using routine mathematical operations or calculations. Between 5 and 19 percent of lessons were taught by teachers who mentioned reasoning mathematically, between 11 and 16 percent of lessons were taught by teachers who mentioned applying mathematics to real world problems, and between 11 and 19 percent of lessons were taught by teachers who mentioned knowing mathematical content. Between 6 and 15 percent of lessons in all of the countries for which reliable estimates could be calculated were taught by teachers who mentioned other process goals, including such processes as acquiring problem solving abilities, meeting external requirements, or reviewing mathematical concepts or problems. Across the various kinds of process goals, no differences were detected among countries on the percentage of lessons taught by teachers who identified that goal.

TABLE 2.5. Percentage of eighth-grade mathematics lessons taught by teachers who identified specific process goals for the videotaped lesson, by country: 1999

Process goal	AU	CZ	HK	NL	SW	US
Using routine operations[2]	40	51	51	42	44	41
Reasoning mathematically[3]	7	9	17	19	5	8
Applying mathematics to real-world problems[4]	12	16	11	16	14	14
Knowing mathematical content[5]	16	11	13	12	18	19
Other process goal[6]	14	6	‡	10	12	15
No process goal identified	10	6	5	‡	7	4

‡Reporting standards not met. Too few cases to be reported.
[1]AU=Australia; CZ=Czech Republic; HK=Hong Kong SAR; NL=Netherlands; SW=Switzerland; and US=United States.
[2]Using routine operations: No differences detected.
[3]Reasoning mathematically: No differences detected.
[4]Applying mathematics to real-world problems: No differences detected.
[5]Knowing mathematical content: No differences detected.
[6]Other process goal: No differences detected.
NOTE. Teachers' responses were coded into one category only. Percentages may not sum to 100 because of rounding and data not reported.
SOURCE: U.S. Department of Education, National Center for Education Statistics, Third International Mathematics and Science Study (TIMSS), Video Study, 1999.

Other Factors Influencing Content in the Videotaped Lessons

Factors other than teachers' learning goals can influence teachers' decisions about what and how they teach. The mathematics teachers were asked to identify whether various documents, guides, or other factors contributed to their decision to teach the content captured in the videotaped lesson. Among the response options were national, state, district, or school curriculum guidelines, external examinations or standardized tests,[4] mandated textbooks, personal comfort or interest, personal assessment of the students' interests or needs, and cooperative work with other teachers or consultants.

Table 2.6 shows that curriculum guidelines reportedly played a major role in teachers' choices about what to teach in Australian (83 percent) and Czech (96 percent) lessons compared to Hong Kong SAR, Dutch, and Swiss lessons. In the teacher questionnaire, curriculum guideline was not specifically defined, leaving it open to respondents' interpretation. However, the intent was to include any document that specified generally or specifically what should be included or covered in the curriculum. External examinations or standardized tests were prominent in the decisions of teachers in 44 percent of the lessons in Hong Kong SAR and 38 percent of U.S. lessons, both larger percentages than in the Netherlands (8 percent). Mandated textbooks played a larger role in Dutch lessons (97 percent) compared to lessons in all the other countries. Teachers' assessment of students' needs had more influence on lessons in Australia, the Czech Republic, Switzerland, and the United States (ranging from 29 to 63 percent) than lessons in the Netherlands (15 percent). Fifty-nine percent of lessons in the Netherlands were taught by teachers who indicated that cooperative work with other teachers was a major influence, a significantly higher percentage compared to all the other countries for which reliable estimates could be calculated.

TABLE 2.6. Percentage of eighth-grade mathematics lessons taught by teachers who reported that various factors played a "major role" in their decision to teach the content in the videotaped lesson, by country: 1999

Factor	AU	CZ	HK	NL	SW[2]	US
Curriculum guidelines[3]	83	96	49	39	59	70
External exams or tests[4]	—	‡	44	8	17	38
Mandated textbook[5]	28	41	64	97	57	40
Teacher's comfort with or interest in the topic[6]	29	28	25	12	23	36
Teacher's assessment of students' interests or needs[7]	41	48	29	15	54	63
Cooperative work with other teachers[8]	26	‡	14	59	12	12

—Not available.
‡Reporting standards not met. Too few cases to be reported.
[1]AU=Australia; CZ=Czech Republic; HK=Hong Kong SAR; NL=Netherlands; SW=Switzerland; and US=United States.
[2]French- and Italian-speaking areas of Switzerland only.
[3]Curriculum guidelines: AU, CZ>HK, NL, SW; CZ>US.
[4]External exams or tests: HK, US>NL; HK>SW.
[5]Mandated textbook: HK>CZ; NL>AU, CZ, HK, SW, US; HK, SW>AU.
[6]Teacher's comfort with or interest in the topic: US>NL.
[7]Teacher's assessment of students' interests or needs: AU, CZ, SW, US>NL; SW, US>HK.
[8]Cooperative work with other teachers: NL>AU, HK, SW, US.
NOTE: Percentages based on mathematics teachers' reports. Percentages do not sum to 100 because more than one category could be selected.
SOURCE: U.S. Department of Education, National Center for Education Statistics, Third International Mathematics and Science Study (TIMSS), Video Study, 1999.

[4]The option "External examinations or standardized tests" was not appropriate for Australia or for the German-speaking area of Switzerland and was excluded from their teacher questionnaires. Analyses of this option exclude Australia and include only the Italian- and French-speaking areas of Switzerland.

Were Lesson Goals Achieved?

A lesson does not always play out as intended. Interruptions, the need to revisit topics, technical difficulties, and other factors may serve as obstacles to conducting the lesson as planned. To give the filmed teachers the opportunity to describe how closely their goals for the lesson matched the outcomes of the lesson, they were asked if they were satisfied that they achieved their stated goals. In all countries, eighth-grade mathematics lessons were taught by teachers who were similarly satisfied that their lessons played out as they had intended (no country differences detected; data not shown in table or figure). At least 83 percent of lessons in all countries were taught by teachers who responded that they were satisfied with their lessons.

Teachers and Current Ideas About Teaching and Learning Mathematics

Several questionnaire items were designed to identify how teachers might have been influenced by current ideas about teaching and learning mathematics. Because "current ideas" might vary according to the policies, values, and goals of each nation's education system, the phrasing of these items was intentionally broad so teachers could interpret each question within the context of their country. First, teachers were asked if they agreed or disagreed that they were familiar with current ideas in mathematics teaching and learning, or if they had no opinion. Figure 2.2 shows that, on average, more Australian, Dutch, Swiss, and U.S. lessons were taught by teachers who agreed they were familiar with current ideas in mathematics teaching and learning compared to Czech and Hong Kong SAR lessons. At least 69 percent of eighth-grade mathematics lessons in Australia, the Netherlands, Switzerland, and the United States were taught by teachers who agreed that they were familiar with current ideas. In contrast, 63 percent of Czech and Hong Kong SAR lessons were taught by teachers who responded that they had no opinion about their familiarity with current ideas.

FIGURE 2.2. Percentage distribution of eighth-grade mathematics lessons according to whether the teachers believed that they were familiar with current ideas in mathematics teaching and learning, by country: 1999

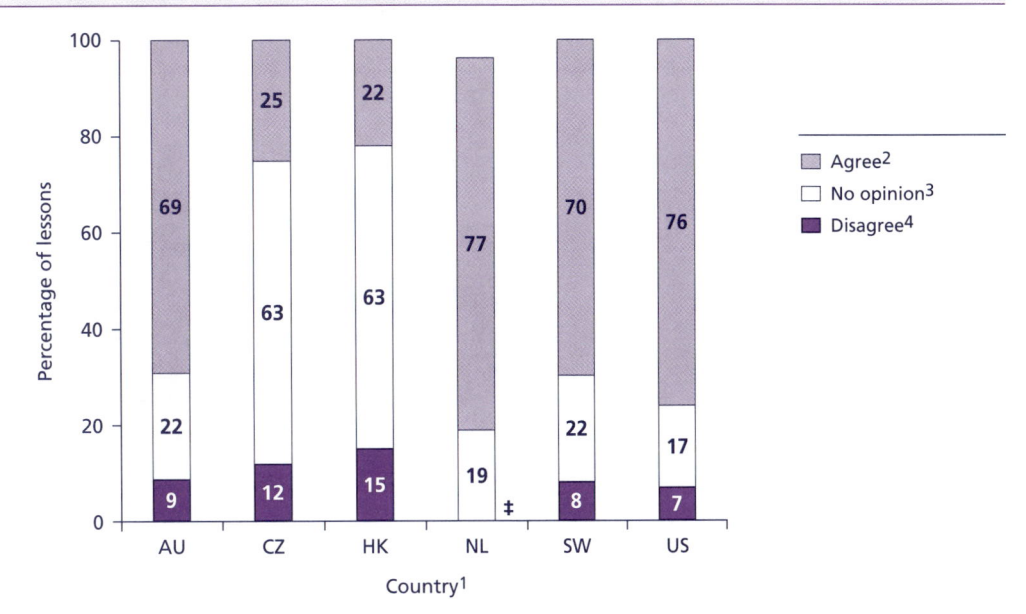

‡Reporting standards not met. Too few cases to be reported.
[1]AU=Australia; CZ=Czech Republic; HK=Hong Kong SAR; NL=Netherlands; SW=Switzerland; and US=United States.
[2]Agree: AU, NL, SW, US>CZ, HK.
[3]No opinion: CZ, HK>AU, NL, SW, US.
[4]Disagree: No differences detected.
NOTE: Percentages may not sum to 100 because of data not reported.
SOURCE: U.S. Department of Education, National Center for Education Statistics, Third International Mathematics and Science Study (TIMSS), Video Study, 1999.

To understand how teachers might have implemented their knowledge of current ideas, they were asked to rate the degree to which the videotaped lesson reflected current ideas about teaching and learning mathematics. Figure 2.3 shows that at least 44 percent of eighth-grade mathematics lessons in all countries except Hong Kong SAR were taught by teachers who believed that their lessons contained a fair amount or a lot of aspects that reflect current ideas. In particular, U.S. lessons were taught by teachers who described their lessons as more consistent with current ideas relative to teachers in all other countries except Australia. On the other hand, more Hong Kong SAR lessons (61 percent) were taught by teachers who reported that the lesson did not reflect current ideas at all compared to Czech, Dutch, and Swiss lessons.

FIGURE 2.3. Percentage distribution of eighth-grade mathematics lessons taught by teachers who rated the extent to which the videotaped lesson was in accord with current ideas about teaching and learning mathematics, by country: 1999

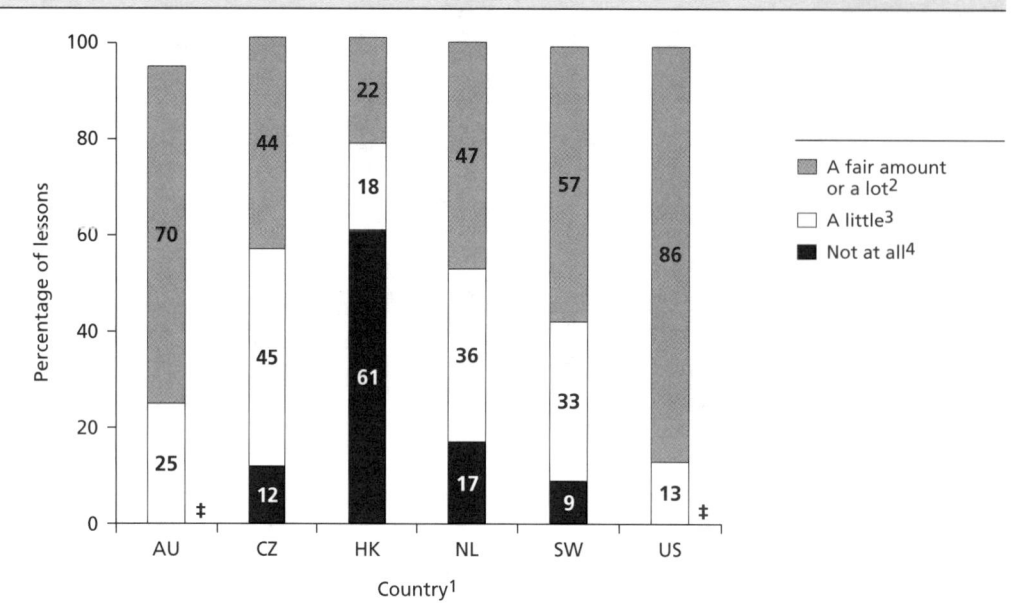

‡Reporting standards not met. Too few cases to be reported.
[1]AU=Australia; CZ=Czech Republic; HK=Hong Kong SAR; NL=Netherlands; SW=Switzerland; and US=United States.
[2]A fair amount or a lot: AU, CZ, NL, SW, US>HK; AU, US>CZ; US>NL, SW.
[3]A little: CZ, NL, SW>US; CZ>HK.
[4]Not at all: HK>CZ, NL, SW.
NOTE: Percentages may not sum to 100 because of rounding and data not reported.
SOURCE: U.S. Department of Education, National Center for Education Statistics, Third International Mathematics and Science Study (TIMSS), Video Study, 1999.

Teachers' Perceptions of the Typicality of the Videotaped Lesson

Several questionnaire items asked teachers to describe how typical the videotaped lesson and their planning for the videotaped lesson were, and to describe the influence of the camera on the lesson. To provide a context for these responses, teachers also were asked about the course of which the videotaped lesson was a part.

Typicality of the Course

Teachers were asked if all eighth-graders in the school took the same mathematics course as the one in the videotaped lesson. Eighth-grade mathematics teachers in the German- and Italian-speaking areas of Switzerland were not asked this question because according to country experts, all students in those schools were required to take the same mathematics course. In this instance, responses for the teachers from the German- and Italian-speaking areas of Switzerland were coded to indicate that all students were required to take the same mathematics course at the school.

More eighth-grade mathematics Czech and Dutch lessons (100 percent), Hong Kong SAR lessons (99 percent), Swiss lessons (86 percent), and Australian lessons (82 percent) were taught by teachers who reported that eighth-grade students were required to take the same mathematics course compared to 25 percent of U.S. lessons. Seventy-five percent of U.S. lessons were taught by teachers who reported that not all students took the same mathematics course in their school (data not shown).

Teachers who indicated that the videotaped mathematics course was not the same course as other eighth-grade students took at the school were asked if they perceived the curriculum to be "more" or "less" challenging or "a typical eighth-grade curriculum" compared to typical eighth-grade mathematics courses in their school. This question was not applicable for the Czech Republic, the Netherlands, and the German- and Italian-speaking areas of Switzerland because the course was required of all students in those schools. In Hong Kong SAR, virtually all teachers (99 percent) indicated that all students took the same course.

Of the 18 percent of Australian eighth-grade mathematics lessons that were taught by teachers who reported that eighth-grade students at their schools were not required to take the same course as the videotaped mathematics course, 43 percent were taught by teachers who identified the videotaped course as more challenging and 46 percent were taught by teachers who identified the course as typical. Of the 75 percent of U.S. eighth-grade mathematics lessons that were taught by teachers who reported that eighth-grade students at their schools were not required to take the same course, 48 percent were taught by teachers who said the videotaped course was more challenging and 49 percent were taught by teachers who said it was typical (data not shown).

The reader should keep in mind that, because differentiation of students into different courses based on ability or interest can occur at either the school level or within the school, teachers' responses to the questionnaire item do not necessarily indicate the extent to which all or most students in a country take the same eighth-grade mathematics course. That is, all students may take the same course at a school that is more or less selective.

Typicality of the Videotaped Lesson

Teachers' judgments of typicality

The eighth-grade mathematics teachers were asked to judge the degree to which the videotaped lesson represented a "typical" mathematics lesson with this group of students in three areas: pedagogy, students' behavior, and content difficulty. With respect to pedagogy, teachers were asked, "How often do you use the teaching methods that are in the videotaped lesson?" Figure 2.4 shows that the majority of lessons were taught by teachers who thought the videotaped lesson portrayed how they "often" or "almost always" taught mathematics. These two response options accounted for between 74 and 97 percent of the responses in each of the six countries. Across the countries, no more than 26 percent of lessons were taught by teachers who reported that they "sometimes" or "seldom" used the teaching methods captured on videotape.

Teaching Mathematics in Seven Countries
Results From the TIMSS 1999 Video Study

FIGURE 2.4. Percentage distribution of eighth-grade mathematics lessons taught by teachers who rated how often they used the teaching methods in the videotaped lesson, by country: 1999

‡Reporting standards not met. Too few cases to be reported.
[1]AU=Australia; CZ=Czech Republic; HK=Hong Kong SAR; NL=Netherlands; SW=Switzerland; and US=United States.
[2]Almost always: No differences detected.
[3]Often: CZ, SW>HK.
[4]Sometimes or seldom: No differences detected.
NOTE: Percentages may not sum to 100 because of rounding and data not reported.
SOURCE: U.S. Department of Education, National Center for Education Statistics, Third International Mathematics and Science Study (TIMSS), Video Study, 1999.

A teacher's ability to conduct a lesson is related, in part, to students' behavior. A second question examining the typicality of the videotaped lesson asked teachers to rate their students' behavior during the lesson. As shown in figure 2.5, at least half of the lessons in each country were taught by teachers who reported that the students behaved about the same as usual except in the Czech Republic (44 percent). Forty-one percent of Czech lessons and no more than 23 percent of lessons across all the other countries were taught by teachers who replied that their students did not behave as well as they usually did. On a follow-up question, in these Czech lessons, the teachers described their students as less active (64 percent), more shy and afraid to give wrong answers (44 percent), or less focused (9 percent) than usual.

| FIGURE 2.5. | Percentage distribution of eighth-grade mathematics lessons by teachers' ratings of their students' behavior in the videotaped lesson, by country: 1999 |

[Bar chart showing percentage of lessons by country with three categories: Better than usual, About the same, Worse than usual]

Country	Worse than usual	About the same	Better than usual
AU	5	73	22
CZ	41	44	14
HK	20	51	30
NL	23	70	7
SW	6	74	21
US	7	73	20

[1]AU=Australia; CZ=Czech Republic; HK=Hong Kong SAR; NL=Netherlands; SW=Switzerland; and US=United States.
[2]Better than usual: HK>NL.
[3]About the same: AU, NL, SW, US>CZ; SW>HK.
[4]Worse than usual: CZ>AU, SW, US.
NOTE: Percentages may not sum to 100 because of rounding.
SOURCE: U.S. Department of Education, National Center for Education Statistics, Third International Mathematics and Science Study (TIMSS), Video Study, 1999.

A third item assessing the lesson typicality explored the difficulty of the mathematics content of the lesson. Teachers were asked if the content for their eighth-grade students was more difficult, less difficult, or about the same level of difficulty as most lessons. Figure 2.6 shows that between 75 and 92 percent of the eighth-grade mathematics lessons in each country were taught by teachers who identified the content level as the same as most lessons. Ten percent of the Czech and Swiss lessons, 7 percent of the Hong Kong SAR lessons, and 6 percent of the Australian and U.S. lessons were taught by teachers who reported that the content of the videotaped lesson was more difficult than usual.

FIGURE 2.6. Percentage distribution of eighth-grade mathematics lessons by teachers' ratings of the difficulty of the lesson content compared to usual, by country: 1999

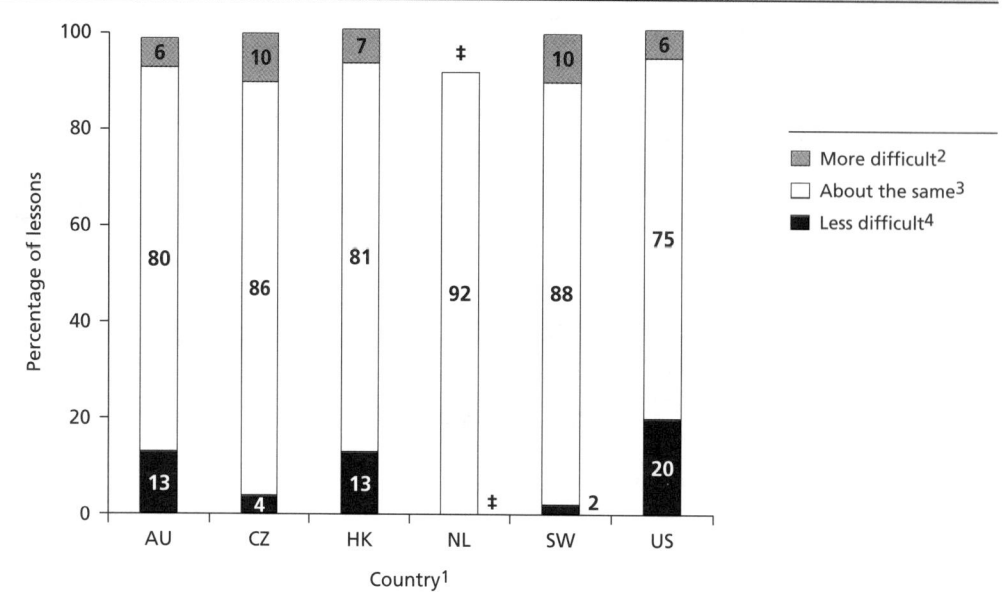

Country	Less difficult	About the same	More difficult
AU	13	80	6
CZ	4	86	10
HK	13	81	7
NL	‡	92	‡
SW	2	88	10
US	20	75	6

‡Reporting standards not met. Too few cases to be reported.
[1]AU=Australia; CZ=Czech Republic; HK=Hong Kong SAR; NL=Netherlands; SW=Switzerland; and US=United States.
[2]More difficult: No differences detected.
[3]About the same: No differences detected.
[4]Less difficult: US>SW.

NOTE: Percentages may not sum to 100 because of rounding and data not reported.
SOURCE: U.S. Department of Education, National Center for Education Statistics, Third International Mathematics and Science Study (TIMSS), Video Study, 1999.

Influence of videotaping

Being videotaped could have affected directly the typicality and quality of the lessons. To check this, teachers were asked specifically about the influence of the video camera in the classroom. They were asked whether the camera caused them to teach a lesson that was worse than usual, about the same, or better than usual. As shown in figure 2.7, between 80 and 91 percent of the eighth-grade mathematics lessons in Australia, the Netherlands, Switzerland, and the United States were taught by teachers who reported that their lesson was "about the same," despite the presence of the video camera. In the Czech Republic and Hong Kong SAR, 38 and 35 percent of lessons, respectively, were taught by teachers who reported that the lesson was worse than usual.[5]

---

[5]The same question was asked of the Japanese teachers in the TIMSS 1995 Video Study. Japanese teachers reported that the videotaped lesson was better than usual in 12 percent of the lessons, the same as usual in 61 percent of the lessons, and worse than usual in 27 percent of the lessons (see Stigler et al. 1999, p. 38).

FIGURE 2.7. Percentage distribution of eighth-grade mathematics lessons taught by teachers who rated the influence of the camera on their teaching of the videotaped lesson, by country: 1999.

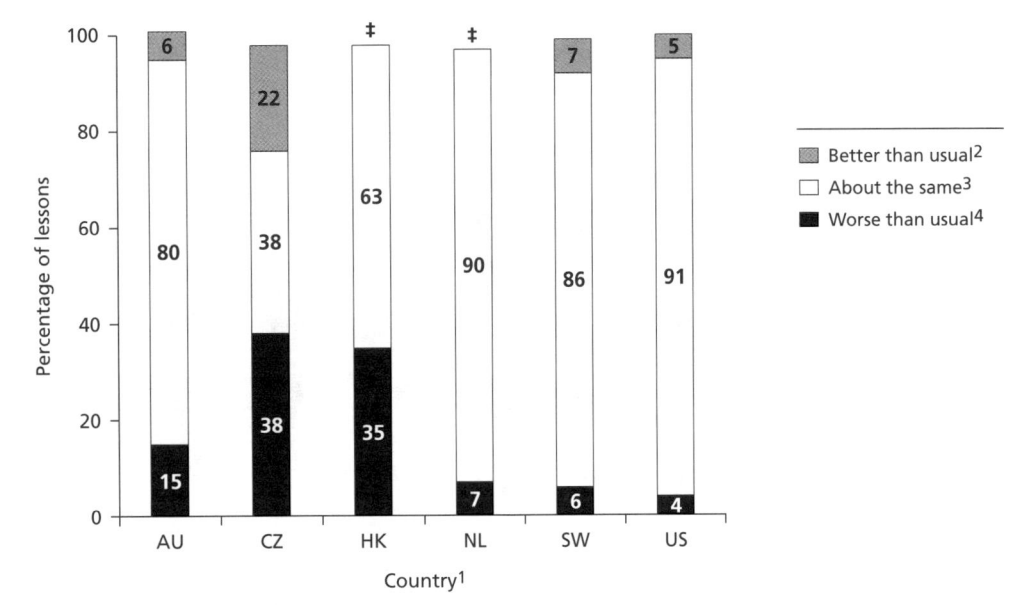

‡Reporting standards not met. Too few cases to be reported.
[1]AU=Australia; CZ=Czech Republic; HK=Hong Kong SAR; NL=Netherlands; SW=Switzerland; and US=United States.
[2]Better than usual: CZ>AU, SW, US.
[3]About the same: AU, HK, NL, SW, US>CZ; NL, SW, US>HK.
[4]Worse than usual: CZ, HK>AU, NL, SW, US.
NOTE: Percentages may not sum to 100 because of rounding and data not reported.
SOURCE: U.S. Department of Education, National Center for Education Statistics, Third International Mathematics and Science Study (TIMSS), Video Study, 1999.

Typicality of planning for the videotaped lesson

In anticipation of being videotaped, the eighth grade mathematics teachers could have invested more effort in planning a lesson, potentially altering how they would normally teach. Teacher reports of how many minutes they spent planning for the videotaped lesson and how many minutes they typically spent planning for a similar mathematics lesson are shown in figure 2.8. Within-country comparisons indicated that, on average, lessons in Australia, the Czech Republic, Hong Kong SAR, and Switzerland were taught by teachers who spent significantly more time planning for the videotaped lesson than usual. On the other hand, no differences between Dutch and U.S. lessons were detected on the average amount of time teachers spent planning for the videotaped lesson compared to the amount of time they usually spent planning for a lesson.

FIGURE 2.8. Length of time averaged over eighth-grade mathematics lessons that teachers reported planning for the videotaped lesson and for similar mathematics lessons, by country: 1999

Country	Videotaped lesson	Similar lessons
AU	39	24
CZ	51	31
HK	46	27
NL	16	12
SW	38	32
US	40	33

[1]AU=Australia; CZ=Czech Republic; HK=Hong Kong SAR; NL=Netherlands; SW=Switzerland; and US=United States.
[2]Videotaped lesson: AU, CZ, HK, SW, US>NL.
[3]Similar lessons: AU, CZ, HK, SW, US>NL.
NOTE: Average length of time calculated as the sum of minutes reported for each lesson divided by the number of lessons within a country.
SOURCE: U.S. Department of Education, National Center for Education Statistics, Third International Mathematics and Science Study (TIMSS), Video Study, 1999.

Fit of the lesson in the curricular sequence

An individual mathematics lesson, ordinarily, is embedded in a sequence designed to teach a particular topic in the curriculum. Lessons that are not part of a sequence might be suspect as atypical lessons conducted especially for the benefit of this study. Therefore, teachers were asked to provide information on whether the videotaped lesson was part of a larger unit or sequence of related lessons, or whether it was a "stand-alone" lesson. Between 92 and 100 percent of the eighth-grade mathematics lessons in all countries were taught by teachers who reported that the videotaped lesson was part of a sequence, with no between-country differences found (data not shown).[6] Excluding the Czech Republic for which there were not enough lessons to calculate a reliable estimate, the percentage of stand-alone lessons ranged from 2 percent in Australia and the Netherlands to 8 percent in Hong Kong SAR and the United States.

If the lesson was part of a unit, the teacher was asked to identify how many lessons were in the entire unit and where the videotaped lesson fell in the sequence (e.g., lesson number 3 out of 5 in the unit). Table 2.7 shows that, on average, the total number of lessons in the larger unit of which the videotaped lesson was a part ranged from 9 to 15. Units in the Czech Republic, averaging 15 lessons per unit, were significantly longer than average units in all other countries except Switzerland. On average, the lessons captured on videotape were located within the middle third of the lessons within the unit.

---

[6]The same question was asked of the Japanese teachers in the TIMSS 1995 Video Study. Ninety-six percent of lessons were taught by a teacher who reported that the videotaped lesson was part of a sequence (Stigler et al. 1999).

TABLE 2.7. Average number of eighth-grade mathematics lessons in unit and placement of the videotaped lesson in unit, by country: 1999

Country	Average number of lessons in unit[1]	Average placement of the videotaped lesson in unit
Australia (AU)	10	5
Czech Republic (CZ)	15	8
Hong Kong SAR (HK)	9	4
Netherlands (NL)	9	5
Switzerland (SW)	12	6
United States (US)	9	5

[1]Average number of lessons in unit: CZ>AU, HK, NL, US.
SOURCE: U.S. Department of Education, National Center for Education Statistics, Third International Mathematics and Science Study (TIMSS), Video Study, 1999.

Summary

In this chapter, teacher questionnaire results were presented in three major areas: teachers' preparation, workload, and learning goals; teachers' awareness of, and implementation of, current ideas of mathematics teaching; and teachers' perceptions of the typicality of the videotaped lesson. Japan was not included in these analyses; results from Japanese teacher questionnaire responses in the TIMSS 1995 Video Study are presented in Stigler et al. (1999).

Some of the key results from this chapter are discussed below. These findings provide a context within which to interpret those presented in the following chapters on the nature of the videotaped lessons.

- Based on teachers' responses to general indicators of teacher preparedness to teach mathematics, such as formal education and certification, the average eighth-grade mathematics lesson in all of the countries seems, at least minimally, to have a teacher who has postsecondary education in mathematics and is certified to teach mathematics at grade 8.

  o Forty-one percent of eighth-grade mathematics lessons in Hong Kong SAR and at least 57 percent of lessons in all the other countries were taught by teachers who reported mathematics or mathematics education as a major field of study in their post-secondary education (table 2.1). When including teachers' reports of minor fields of study in addition to a major field of study, 58 percent of lessons in Switzerland and at least 83 percent of lessons in all the other countries were taught by teachers who identified mathematics or mathematics education as a major or a minor field of study.

  o The percentage of lessons that were taught by teachers who were certified to teach eighth-grade mathematics ranged from 48 (Switzerland) to 91 (the Netherlands) (table 2.2).

- The eighth-grade mathematics lessons were taught by teachers who reported spending on average from 36 hours per week (in Australia) to 42 hours per week (in the Czech Republic, Switzerland, and the United States) on school-related work activities (table 2.4). Lessons were taught by teachers who spent more hours teaching mathematics on average in the Netherlands (20 hours) and the United States (18 hours) than in the other countries, and more hours spent teaching other subjects in Switzerland (13 hours) than in the other countries.

- Relatively few country differences were found in the kinds of learning goals the eighth-grade mathematics teachers identified for the videotaped lesson. In particular, no country differences were found on the percentage of lessons taught by teachers who identified the different types of process goals: using routine operations (40–51 percent); reasoning mathematically (5–19 percent); applying mathematics to real-world problems (11–16 percent); and knowing mathematical content (11–19 percent) (table 2.5).

- In addition to teachers' learning goals, the content of the videotaped lesson was influenced by a variety of factors, with different emphases found in different countries (table 2.6). More eighth-grade mathematics lessons in the Netherlands were taught by teachers who, compared with their peers, indicated that textbooks and cooperative work with other teachers played a major role in their choices about what to teach. More lessons in Australia and the Czech Republic were taught by teachers whose choices were strongly influenced by curriculum guidelines compared to Hong Kong SAR, Dutch, and Swiss lessons. In Hong Kong SAR and the United States, a larger percentage of the lessons than in the Netherlands were taught by teachers who cited external exams as playing a major role in lesson content decisions.

- A larger percentage of Australian, Dutch, Swiss, and U.S. eighth-grade mathematics lessons were taught by teachers who reported that they were familiar with "current ideas" of teaching mathematics than lessons in the Czech Republic and Hong Kong SAR (figure 2.2). Eighty-three percent or more of lessons in all countries, except Hong Kong SAR (40 percent), were taught by teachers who believed the videotaped lesson was, at least "a little," in accord with current ideas (figure 2.3). U.S. teachers expressed greater confidence that, on average, their lesson was in accord with current ideas than teachers in all other countries except Australia.

- The videotaped lesson, as perceived by teachers of the eighth-grade mathematics lessons, generally provided a picture of everyday classroom instruction with regard to teaching methods (figure 2.4), content difficulty (figure 2.6), its fit within a curriculum unit (table 2.7), and, with the exception of the Czech Republic, students' behavior (figure 2.5). In the Netherlands and the United States, there were no differences detected between the amount of planning teachers did for the videotaped lesson compared to their usual planning time, while in the other countries the teachers spent significantly more time planning for the videotaped lesson than usual (figure 2.8).

The results presented in this chapter suggest a rather complicated patchwork of similarities and differences among eighth-grade mathematics lessons in different countries. Differences exist, for example, in the formal preparation for teaching, in the weekly workloads, and in the awareness of "current ideas" of teaching mathematics. Different subsets of the countries, however, are similar and different on a variety of variables so that simple patterns are difficult to discern. Moreover, lessons in all countries show considerable similarity on some variables, such as the learning goals for the videotaped lesson.

Perhaps the most important finding for this analysis is that most eighth-grade mathematics lessons in most countries were taught by teachers who considered the videotaped lesson to be typical of their teaching. This adds credibility to the findings reported in the following chapters. Although questionnaires provide contextual information that can be referenced when examining the videotaped lessons, the substantive contributions of this study come from the videotaped lessons themselves. As noted in chapter 1, there are considerable advantages of video data over self-reports of teaching, and the goal of the following chapters is to exploit these advantages to describe and compare mathematics teaching in seven countries.

CHAPTER 3
The Structure of Lessons

Students' opportunities to learn mathematics during classroom lessons can be influenced by a variety of factors, including the knowledge and skills they already possess, as well as the activities in which they engage during the lesson (National Research Council 1999, 2001a). Eighth-grade students bring to their mathematics lessons experiences from their home environments as well as from earlier grades. Such experiences, along with other factors such as curriculum guidelines and a mandated textbook (see chapter 2, table 2.6), might impact how their teachers choose to conduct their classrooms. The implication for this study is that the videotaped lessons only capture a part of the story regarding students' learning opportunities in each country.

Nonetheless, the videotaped lessons present a wealth of opportunities for examining eighth-grade mathematics teaching practices, as discussed in chapter 1. Based on earlier experience (Stigler et al. 1999), these lessons were analyzed to address three broad questions: (1) how was the lesson environment organized (e.g., what kind of activities took place, and what was the purpose of these activities)? (2) what kind of mathematics content was studied? and (3) how was the content studied? These three questions form the basic organizing principle for the next three chapters of this report.

The danger of separating information into categories is that it can give the mistaken impression that the categories contribute independently to students' learning opportunities. Rather, it seems likely that these categories interact to shape the learning opportunities of students to varying degrees. The reader is cautioned that one cannot assume that each of the pieces of data presented contribute to increased learning opportunities, or that each is independent and can or should be manipulated separately to improve the educational experiences of students.

This chapter presents information on the way in which the mathematics lesson environments in each country were organized. The organization of the lesson may constrain both the mathematics content that is taught and the way that content is taught. Furthermore, examining the lesson organization itself reveals important similarities and differences in eighth-grade mathematics classrooms across countries.

Teachers can organize eighth-grade mathematics lessons in various ways, shaping them for particular purposes and for particular groups of students. For example, a teacher might engage students in private work at their desks for most of a lesson rather than engage them in whole-class discussion. This interaction structure is likely to influence the kind of mathematical work that might be done, who might do the work, and the kind of learning experiences that students might have (Brophy 1999). Similarly, a teacher might devote most of a lesson to reviewing previously taught material rather than introducing new material, thereby shaping further the learning opportunities for students.

This chapter sets the stage for the remaining chapters by describing the kinds of broad organizational elements that were prevalent in eighth-grade mathematics lessons across the participating countries. The following elements of lesson organization were examined:

- The amount of time spent studying mathematics during classroom lessons;

- The main type of activity used to study mathematics in classrooms—solving mathematical problems;

- Ways in which lessons were divided among reviewing old material, introducing new material, and practicing new material;

- The grouping structures used to study mathematics—whole-class public discussions, private independent work, and combinations;

- The role of homework; and

- Ways in which key ideas were clarified and lesson flow was enhanced or interrupted.

Together, these elements of lesson organization contribute to the shape of the learning environment for students. The research literature does not definitively suggest a preferred combination of these elements, or a right or wrong way of arranging them, so the data are not presented to portray which country creates the "right" environment for students. But, as will be seen, different countries make different choices and the comparisons provide a chance for educators to examine whether the choices they are making are aligned with their learning goals.

The Length of Lessons

The length of a mathematics lesson provides the most basic element of lesson organization. Although amount of time does not, by itself, account for students' learning opportunities, it is a necessary ingredient for learning (National Research Council 1999). So, the amount of time devoted to formal study of mathematics is a good starting point for describing classroom lessons; how the teachers and students filled in the time with mathematical work will become apparent over the course of this chapter and the following two chapters.

To ensure that the eighth-grade mathematics lessons filmed for this study were captured in their entirety, the data collection protocol called for videographers to turn on their cameras well before the lesson started and continue filming even after the lesson ended. To calculate the length of a mathematics lesson, decisions had to be made about when a lesson began and ended. The beginning of the lesson was defined as the point when the teacher first engaged in talk intended for the entire class. The end of a lesson was marked by the teacher's final talk intended for the entire class, which sometimes included concluding or summary remarks by the teacher. When students worked independently and the teacher did not close the lesson with a public statement, the end of lesson was marked when the bell rang, or when most students packed up their materials and left the classroom.

These definitions reflect a deliberate intention to capture the length of the entire class period, and not just the mathematics portion of the lesson. In many cases, lessons began or ended with non-mathematical activities. These activities were included in the lesson and later marked as "non-mathematical segments." It is therefore the case that the recorded time for a given lesson might not correspond exactly to the officially designated length of that class period.

The lesson duration mean, median, range, and standard deviation for each country are displayed in table 3.1.

TABLE 3.1. Mean, median, range, and standard deviation (in minutes) of the duration of eighth-grade mathematics lessons, by country: 1999

Country	Mean[2]	Median	Range	Standard deviation[3]
Australia (AU)	47	45	28–90	13
Czech Republic (CZ)	45	45	41–50	1
Hong Kong SAR (HK)	41	36	26–91	14
Japan[1] (JP)	50	50	45–55	2
Netherlands (NL)	45	45	35–100	7
Switzerland (SW)	46	45	39–65	3
United States (US)	51	46	33–119	17

[1]Japanese mathematics data were collected in 1995.
[2]Mean: AU, CZ, JP, NL, SW, US>HK; JP>CZ, NL, SW; US>CZ.
[3]Standard deviation: AU, HK>CZ, JP; US>CZ, JP, NL, SW.
NOTE: Mean is calculated as the sum of the number of minutes of each lesson divided by the number of lessons within a country. For each country, median is calculated as the number of minutes below which 50 percent of the lessons fall. Range describes the lowest number of minutes and the highest number of minutes observed within a country.
SOURCE: U.S. Department of Education, National Center for Education Statistics, Third International Mathematics and Science Study (TIMSS), Video Study, 1999.

There are two features of lesson length that are immediately apparent: eighth-grade mathematics lessons across the countries had mean lengths between 41 and 51 minutes, and there was a large range in lesson length in some countries. With regard to mean length, Hong Kong SAR eighth-grade mathematics lessons were shorter than those of all the other countries, and Japanese lessons were longer than those of four countries. Because of the large variations, however, the median length is probably the best measure for gauging the length of a typical lesson. The large range of lengths in some countries was due, in part, to what some countries call "double lessons," lessons in which two traditional instructional periods are joined.

Figure 3.1 displays the distribution of lesson durations for each country. This figure shows graphically the clustering of lesson lengths at around 45 minutes for all the countries except Japan and Hong Kong SAR. The figure provides a more detailed look at the variation in lesson length. Whereas table 3.1 showed that the ranges in lesson duration differed widely, the box and whisker plots in figure 3.1 reveal that the majority of lessons in all countries except Australia fall within a narrower range. The figure also indicates the lessons that were extremes or outliers in terms of duration.

FIGURE 3.1. Box and whisker plots showing the distribution of eighth-grade mathematics lesson durations, by country: 1999

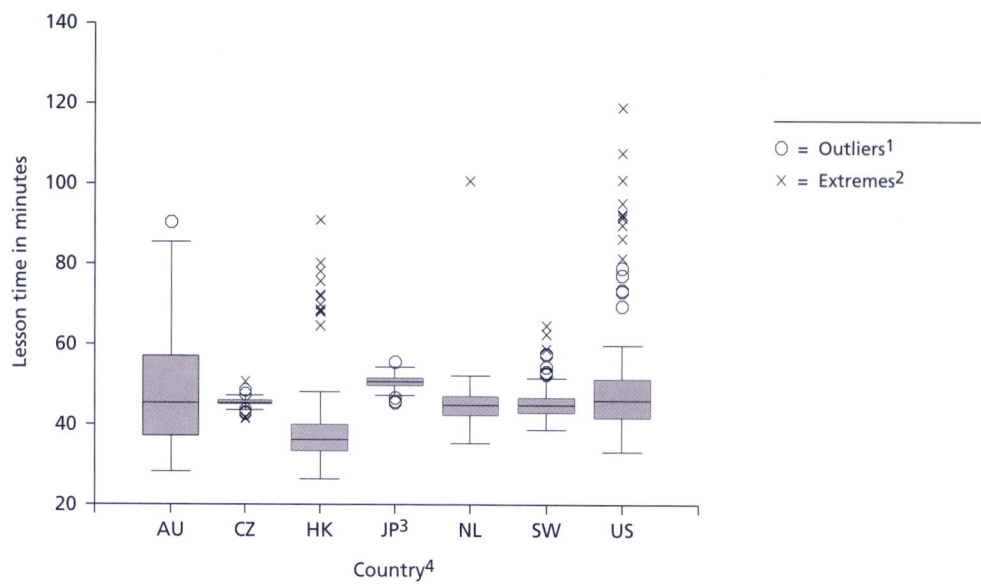

[1]Outliers are values from 1.5 to 3.0 box lengths from the upper or lower edge of the box.
[2]Extremes are values greater than 3.0 box lengths from the upper or lower edge of the box.
[3]Japanese mathematics data were collected in 1995.
[4]AU=Australia; CZ=Czech Republic; HK=Hong Kong SAR; JP=Japan; NL=Netherlands; SW=Switzerland; and US=United States.
NOTE: The shaded box represents the interquartile range, containing 50 percent of the lessons. The lines extending from the box indicate the highest and lowest values, excluding outliers and extremes. The horizontal line within the box indicates the median lesson time (half of the numbers fall above or below this value).
SOURCE: U.S. Department of Education, National Center for Education Statistics, Third International Mathematics and Science Study (TIMSS), Video Study, 1999.

The Amount of Time Spent Studying Mathematics

Although lesson length provides the boundaries of possible instruction time, the measure of most interest is the time actually spent working on mathematics. Because lesson time can be spent on other things, such as chatting about a musical concert the students attended the night before, it is important to mark the segments of the lesson devoted to mathematical work.

Broadly speaking, there are several ways in which time can be spent during instructional periods:

- *Mathematical work:* Time spent on mathematical content presented either through a mathematical problem or outside the context of a problem, e.g., talking or reading about mathematical ideas, solving mathematical problems, practicing mathematical procedures, or memorizing mathematical definitions and rules.

- *Mathematical organization:* At least 30 continuous seconds devoted to preparing materials or discussing information related to mathematics, but not qualifying as mathematical work, e.g., distributing materials used to solve problems, discussing the grading scheme to be used on a test, distributing a homework assignment [Video clip example 3.1].

- *Non-mathematical work:* At least 30 continuous seconds devoted to non-mathematical content, e.g., talking about a social function, disciplining a student while other students wait, or listening to school announcements on a public-address system [Video clip example 3.2].

To code every minute of every lesson, two additional categories were needed:

- *Break:* Time during the lesson, or between double lessons, that teachers designated as an official break for students.

- *Technical problem:* Time during the lesson when there was a technical problem with the video (such as lack of audio) that prevented members of the international coding team from making confident coding decisions about the segment.

The five types of lesson segments were mutually exclusive and exhaustive. Every second of every eighth-grade mathematics lesson was coded into just one of the five types. To aid in coding, it was decided that mathematics organization and non-mathematical work segments that were less than 30 seconds would not be noted or coded as such. For example, an exchange between a teacher and student during mathematical work that focused on disciplining a student and lasted less than 30 seconds was not noted.

Figure 3.2 displays the average percentage of lesson time devoted to mathematical work, mathematical organization, and non-mathematical work, for each country. The categories "break" and "technical problem" together accounted for less than 1 percent of time in each country and are therefore not shown in the figure.

FIGURE 3.2. Average percentage of eighth-grade mathematics lesson time devoted to mathematical work, mathematical organization, and non-mathematical work, by country: 1999

Country	Mathematical work	Mathematical organization	Non-mathematical work
AU	95	4	1
CZ	98	1	1
HK	97	2	1
JP[1]	98	1	1
NL	95	2	2
SW	97	2	1
US	95	3	1

[1] Japanese mathematics data were collected in 1995.
[2] AU=Australia; CZ=Czech Republic; HK=Hong Kong SAR; JP=Japan; NL=Netherlands; SW=Switzerland; and US=United States.
[3] Non-mathematical work: NL>CZ, HK, JP, SW.
[4] Mathematical organization: AU>CZ, HK, JP, SW; NL>CZ; US>CZ, JP, SW.
[5] Mathematical work: CZ, JP>AU, NL, US; SW>AU, US, NL; HK>AU.

NOTE: Percentages may not sum to 100 because of rounding. The tests for significance take into account the standard error for the reported differences. Thus, a difference between averages of two countries may be significant while the same difference between two other countries may not be significant. For each country, the average percentage was calculated as the sum of the percentage within each lesson, divided by the number of lessons.

SOURCE: U.S. Department of Education, National Center for Education Statistics, Third International Mathematics and Science Study (TIMSS), Video Study, 1999.

In all the countries, on average, at least 95 percent of eighth-graders' lesson time focused on mathematical work. The percentages ranged from 95 in Australia, the Netherlands, and the United States, to 98 in the Czech Republic and Japan. Multiplying these percentages by the median lesson time for each country (see table 3.1) yields an estimated median time (in minutes) spent on mathematical work: Australia: 43; the Czech Republic: 44; Hong Kong SAR: 35; Japan: 49; the Netherlands: 43; Switzerland, 44; the United States: 44. In summary, in all the countries most lessons focused almost entirely on mathematical work.

Comparing mathematical work time across countries based solely on information from a single lesson can be misleading. One lesson may not accurately indicate how much time students spend studying mathematics in school over the course of a week or a year. Countries differ in the number of lessons conducted per week and the number of school weeks per year. By using the estimated median work time per lesson, however, it is possible to estimate the amount of time eighth-graders in each country might spend studying mathematics in school during the week and during the entire school year.

Based on estimates of the number of eighth-grade mathematics lessons per week and per year in each country provided by the TIMSS 1999 Video Study National Research Coordinators,[1] estimates were calculated for the median total time spent in mathematical work per week and per year for each country except Switzerland. The three language regions in Switzerland have different school calendars and it was deemed inappropriate to develop one estimate to represent all three regions. Table 3.2 displays the results.

TABLE 3.2. Estimated median time spent in mathematical work per week and per year in eighth grade, by country: 1999

Country	Estimated median time in mathematical work per week (in minutes)	Estimated median time in mathematical work per year (in hours)
Australia (AU)	174	113
Czech Republic (CZ)	179	90
Hong Kong SAR (HK)	175	105
Japan[1] (JP)	200	116
Netherlands (NL)	127	84
United States (US)	179	107

[1]Japanese mathematics data were collected in 1995. The estimate for Japan is based on the average number of mathematics lessons per week and per year in 1995.
NOTE: Based on estimates of the number of eighth-grade mathematics lessons per week and per year in each country provided by the TIMSS 1999 Video Study National Research Coordinators. Estimates were calculated for the median total time spent in mathematical work per week and per year.
SOURCE: U.S. Department of Education, National Center for Education Statistics, Third International Mathematics and Science Study (TIMSS), Video Study, 1999.

The estimates in table 3.2 are calculated based on data from various sources. These estimates should therefore be considered indicative rather than definitive. Moreover, these estimates are limited to in-school instruction and may not accurately reflect the total amount of instruction

---

[1]The estimates provided by the National Research Coordinators may differ from estimates from other mathematics educators or teachers in these countries.

that students receive in other settings.[2] For this reason, it was deemed inappropriate to compare them statistically. Nonetheless, and as suggested above, the entries in table 3.2 indicate that it might be inappropriate to presume that the individual lesson duration describes the relative time spent by students in each country studying mathematics in school. For example, whereas eighth-grade mathematics lessons in Hong Kong SAR had the shortest mean duration of all the countries (see table 3.1), when taking into account the number of lessons per week and per year, Hong Kong SAR now lies in about the middle range among the countries (see table 3.2).

The Role of Mathematical Problems

Time Spent on Problems and Non-Problems

What did mathematical work time consist of in these eighth-grade lessons? While reviewing the videotapes that were arriving from each country, it became apparent that a considerable portion of lesson time in every country was spent solving mathematics problems. During the remaining time, the teacher might, for example, give a brief lecture. The question was whether "mathematics problem" could be defined with enough clarity and precision to enable reliable coding. There was no precedent from other large-scale studies of mathematics teaching that could be followed.[3] Eventually, a reliable code was developed for mathematics problem so that mathematical work time could be divided into problem and non-problem segments. The definitions of these activities are summarized below (for complete definitions of these codes, see the forthcoming technical report, Jacobs et al.):

- *Working on problems:* Problems were defined as events that contained a statement asking for some unknown information that could be determined by applying a mathematical operation. Problems varied greatly in length and complexity, ranging from routine exercises to challenging problems. Although problems could be relatively undemanding, they needed to require some degree of thought by eighth-grade students. Simple questions asking for immediately accessible information did not count as problems. Mathematical operations of the following kinds were common:

  - Adding, subtracting, multiplying, and dividing whole numbers, decimals, fractions, percents, and algebraic expressions;
  - Solving equations;
  - Measuring lines, areas, volumes, angles;
  - Plotting or reading graphs; and
  - Applying formulas to solve real-life problems.

---

[2] Across the countries participating in the study, there are various options available to students to obtain additional instruction or study time related to school subject matter. For example, students may have access to after-school programs, tutoring services, parental assistance or study groups, among other possibilities.

[3] Neubrand (forthcoming) defined and coded mathematics problems in a secondary analysis of a subset of the TIMSS 1995 video sample, but new definitions were needed for the TIMSS 1999 video sample in order to code reliably mathematics problems in the lessons from seven countries.

- *Non-problem segments:* A non-problem segment was defined as mathematical work outside the context of a problem [Video clip example 3.3]. Without presenting a problem statement, teachers (or students) sometimes engaged in:

    - Presenting mathematical definitions or concepts and describing their mathematical origins;
    - Giving an historical account of a mathematical idea or object;
    - Relating mathematics to situations in the real world;
    - Pointing out relationships among ideas in this lesson and previous lessons;
    - Providing an overview or a summary of the major points of the lesson; and
    - Playing mathematical games that did not involve solving mathematical problems (e.g., a word search for mathematical terms).

Figure 3.3 displays the average percentage of lesson time devoted to problem and non-problem segments.

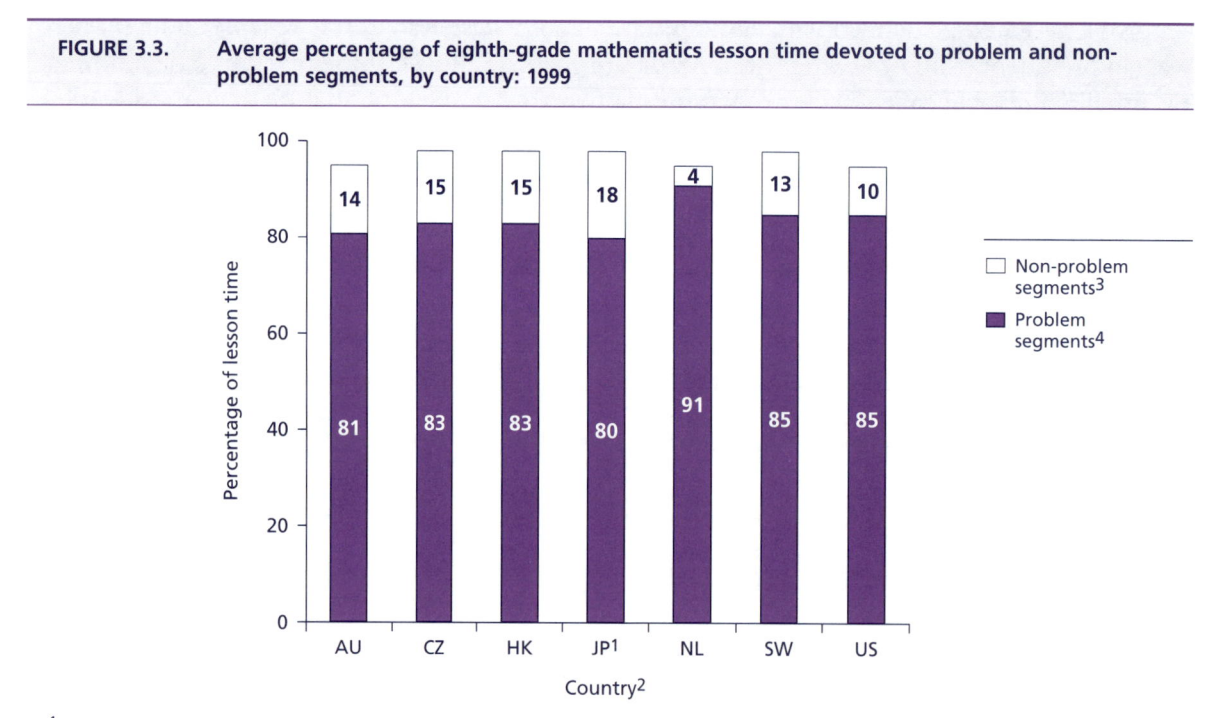

FIGURE 3.3. Average percentage of eighth-grade mathematics lesson time devoted to problem and non-problem segments, by country: 1999

Country	Problem segments	Non-problem segments
AU	81	14
CZ	83	15
HK	83	15
JP[1]	80	18
NL	91	4
SW	85	13
US	85	10

[1]Japanese mathematics data were collected in 1995.
[2]AU=Australia; CZ=Czech Republic; HK=Hong Kong SAR; JP=Japan; NL=Netherlands; SW=Switzerland; and US=United States.
[3]Non-problem segments: AU, CZ, HK, JP, SW, US>NL.
[4]Problem segments: NL>AU, CZ, HK, JP, SW.
NOTE: Percentages sum to average percentage of lesson time devoted to mathematical work per country (see figure 3.2). For each country, average percentage was calculated as the sum of the percentage within each lesson, divided by the number of lessons. The tests for significance take into account the standard error for the reported differences. Thus, a difference between averages of two countries may be significant while the same difference between two other countries may not be significant.
SOURCE: U.S. Department of Education, National Center for Education Statistics, Third International Mathematics and Science Study (TIMSS), Video Study, 1999.

As shown in figure 3.3, in all the countries at least 80 percent of eighth-graders' lesson time, on average, was spent solving problems. A greater percentage of lesson time was spent on mathematical problems in the Netherlands (91 percent) than in all the other countries except the

United States. Conversely, less time was spent on non-problem segments in the Netherlands (4 percent) than in all the other countries.

Independent, Concurrent, and Answered-Only Problems

Because solving problems made up such a large part of eighth-grade students' mathematical work in all the countries, a full description of classroom lessons requires descriptions of the nature of these problems and the way in which they were worked on. In this section, the problems are characterized in terms of the role they played in the lesson organization (whether they were assigned as a set and worked on independently, as homework, as individual problems discussed during class, etc.). Chapter 4 continues the analysis by considering the content of the problems, and chapter 5 focuses on the way in which problems were worked on during the lesson.

Mathematical problems were treated in three different ways or, said another way, played three different roles during the lessons:

- *Independent problems:* Presented as single problems and worked on for a clearly definable period of time [Video clip example 3.4]. These problems might have been solved publicly—as a whole class—or they might have contained a private work phase when students worked on them individually or in small groups.

- *Concurrent problems:* Presented as a set of problems, usually as an assignment from a worksheet or the textbook, to be worked on privately [Video clip example 3.5]. Some of these problems might have eventually been discussed publicly as a whole class. Because they were assigned as a group and worked on privately, it was not possible to determine how long students spent working on any individual problem of this kind.

- *Answered-only problems:* Most often from homework or an earlier test, these problems had already been completed prior to the lesson, and only their answers were shared [Video clip example 3.6]. They included no public discussion of a solution procedure and no time in which students worked on them privately.

It was important to distinguish among the problem types because they can provide different experiences for students. For example, working on a single problem with the whole class can be a different experience from working on a set of problems individually or in small groups, which can be different still from hearing answers to problems completed as homework. More than that, separating out the independent problems, for which it was possible to mark beginning and ending times, allowed further analyses of the nature of these problems.

Table 3.3 displays the average number of independent and answered-only problems per eighth-grade mathematics lesson. The number of concurrent problems per lesson is not reported because such a number is difficult to interpret. Concurrent problems were assigned as a group to be worked on privately. Sometimes the problems were worked during class and sometimes outside of class. Sometimes the problems were to be completed for the next lesson and sometimes the assignment was for an entire week. Consequently, the number of concurrent problems assigned during a lesson provided little reliable information about what happened during the lesson.

Figure 3.4 shows the percentage of eighth-grade mathematics lesson time devoted to the different problem types. As noted above, although it was often unclear how many concurrent problems were actually worked on during the lesson, it was possible to accurately determine the proportion of lesson time devoted to solving concurrent problems. When considered together, table 3.3 and figure 3.4 provide a snapshot of the roles that mathematics problems played in the lessons within and across countries.

TABLE 3.3. Average number of independent and answered-only problems per eighth-grade mathematic lesson, by country: 1999

Country	Independent problems[2]	Answered-only problems[3]
Australia (AU)	7	1
Czech Republic (CZ)	13	#
Hong Kong SAR (HK)	7	#
Japan[1] (JP)	3	‡
Netherlands (NL)	8	2
Switzerland (SW)	5	3
United States (US)	10	5

#Rounds to zero.
‡Reporting standards not met. Too few cases to be reported.
[1]Japanese mathematics data were collected in 1995.
[2]Independent problems: CZ>HK, JP, NL, SW; HK, US>JP, SW; SW, NL>JP.
[3]Answered-only problems: US>CZ, HK.
NOTE: Independent problems were presented as single problems and worked on for a clearly definable period of time. Answered-only problems had already been completed prior to the lesson, and only their answers were shared.
The tests for significance take into account the standard error for the reported differences. Thus, a difference between averages of two countries may be significant while the same difference between two other countries may not be significant.
SOURCE: U.S. Department of Education, National Center for Education Statistics, Third International Mathematics and Science Study (TIMSS), Video Study, 1999.

Chapter 3
The Structure of Lessons

FIGURE 3.4. Average percentage of eighth-grade mathematics lesson time devoted to independent problems, concurrent problems, and answered-only problems, by country: 1999

Country	Independent problems	Concurrent problems	Answered-only problems
AU	26	54	#
CZ	52	31	#
HK	49	33	#
JP[1]	64	16	‡
NL	29	61	1
SW	31	53	1
US	51	32	3

#Rounds to zero.
‡Reporting standards not met. Too few cases to be reported.
[1]Japanese mathematics data were collected in 1995.
[2]AU=Australia; CZ=Czech Republic; HK=Hong Kong SAR; JP=Japan; NL=Netherlands; SW=Switzerland; and US=United States.
[3]Answered-only problems: US>AU, CZ, HK.
[4]Concurrent problems: AU, NL, SW>CZ, HK, JP, US.
[5]Independent problems: CZ, HK, JP, US>AU, NL, SW.

NOTE: Independent problems were presented as single problems and worked on for a clearly definable period of time. Answered-only problems had already been completed prior to the lesson, and only their answers were shared. Concurrent problems were presented as a set of problems to be worked on privately. For each country, average percentage was calculated as the sum of the percentage within each lesson, divided by the number of lessons. Percentages sum to average percentage of lesson time devoted to problem segments per country (see figure 3.3), although in some cases they do not because of rounding or data not reported.
SOURCE: U.S. Department of Education, National Center for Education Statistics, Third International Mathematics and Science Study (TIMSS), Video Study, 1999.

In the Czech Republic, more independent problems were worked on (13) per lesson, on average, than in Hong Kong SAR, Japan, the Netherlands, and Switzerland (see table 3.3). In Japanese lessons, an average of 3 independent problems were worked on per lesson, which is significantly fewer than all the other countries except Australia.

Answered-only problems occurred at least occasionally in all the countries where reliable estimates could be calculated. This type of problem was more common in the United States (averaging 5 per lesson) than in the Czech Republic and Hong Kong SAR.

Figure 3.4 indicates that part of the time in the eighth-grade mathematics lessons in the participating countries was spent solving independent problems and part of the time was spent working on concurrent problems, although in somewhat different proportions. The Czech Republic, Hong Kong SAR, Japan, and the United States devoted a greater percentage of lesson time on average to independent problems than the other three countries. Conversely, students in Australia, the Netherlands, and Switzerland spent more time on average than students in the other countries working on concurrent problems.

A similar picture emerges from within-country comparisons of time spent on independent and concurrent problems. These analyses show that, on average, in the Czech Republic, Hong Kong

SAR, Japan, and the United States more time was devoted to solving independent problems compared to concurrent problems, whereas in Australia, the Netherlands, and Switzerland more time was devoted to solving concurrent problems compared to independent problems (data not shown in table or figure).

Time Spent per Problem

As noted earlier, it was possible to examine independent problems more carefully than concurrent problems because the exact time spent working on each problem could be calculated. This allowed an analysis of the average time spent per independent problem as well as further analyses of the nature of the work that occurred during this time. Various measures of time are presented in this section; descriptions of the content and nature of the activity that occurred while solving the independent problems are presented in later chapters.

Figure 3.5 shows the number of minutes, on average, devoted to each independent problem in a lesson in each country. More time per problem could mean that the problems were more challenging, that the class spent more time discussing the problem, or simply that the teacher allowed more time for students to solve the problem.

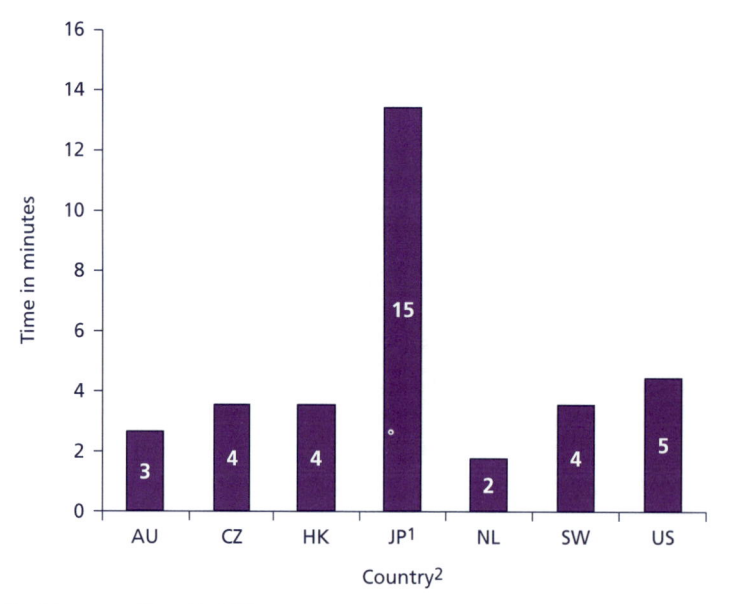

FIGURE 3.5. Average time per independent problem per eighth-grade mathematics lesson (in minutes), by country: 1999

Country	Time in minutes
AU	3
CZ	4
HK	4
JP[1]	15
NL	2
SW	4
US	5

[1] Japanese mathematics data were collected in 1995.
[2] AU=Australia; CZ=Czech Republic; HK=Hong Kong SAR; JP=Japan; NL=Netherlands; SW=Switzerland; and US=United States.
NOTE: CZ, HK, SW>NL; JP>AU, CZ, HK, NL, SW, US. The tests for significance take into account the standard error for the reported differences. Thus, a difference between averages of two countries may be significant while the same difference between two other countries may not be significant.
SOURCE: U.S. Department of Education, National Center for Education Statistics, Third International Mathematics and Science Study (TIMSS), Video Study, 1999.

As the figure shows, in Japanese eighth-grade mathematics lessons, on average, more time was spent working on each independent mathematics problem compared to lessons from all the other countries. Whereas in the other countries, on average, between 2 and 5 minutes was spent working on an independent problem, in Japan, on average, 15 minutes was spent on each independent problem. As shown earlier, Japan also devoted about two-thirds of the lesson time, on average, to independent problems (figure 3.4, 64 percent). From these two findings, one could conclude that Japanese students spent most of their time during mathematics lessons working on a few, independent mathematics problems.

The number of independent problems and time per independent problem were calculated by averaging across lessons. Naturally, there are variations among lessons with regard to the average time spent per independent problem. To get a sense of the variation within countries, it is useful to look at a box and whisker plot showing the distribution of lessons with regard to average time spent on each independent problem. Figure 3.6 shows this variation within each country.

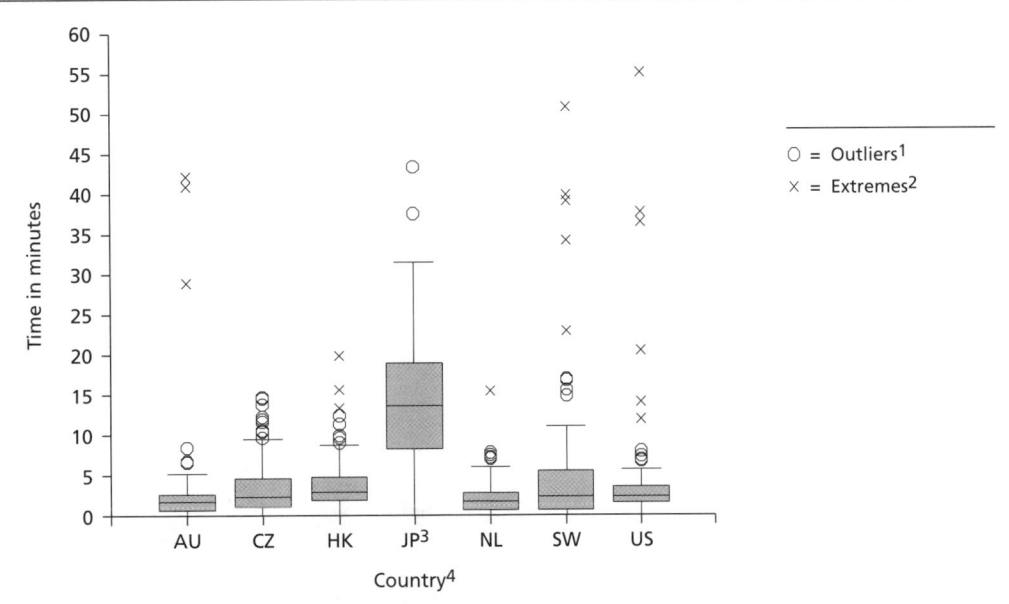

FIGURE 3.6. Box and whisker plots showing the distribution of eighth-grade mathematics lessons based on average length of independent problems, by country: 1999

[1]Outliers are values from 1.5 to 3.0 box lengths from the upper or lower edge of the box.
[2]Extremes are values greater than 3.0 box lengths from the upper or lower edge of the box.
[3]Japanese mathematics data were collected in 1995.
[4]AU=Australia; CZ=Czech Republic; HK=Hong Kong SAR; JP=Japan; NL=Netherlands; SW=Switzerland; and US=United States.
NOTE: The shaded box represents the interquartile range, containing 50 percent of the lessons. The lines extending from the box indicate the highest and lowest values, excluding outliers and extremes. The horizontal line within the box indicates the median.
SOURCE: U.S. Department of Education, National Center for Education Statistics, Third International Mathematics and Science Study (TIMSS), Video Study, 1999.

As evident in figure 3.5, the average length of time per independent problem is higher in Japan than the other countries. However, figure 3.6 shows that the majority of lessons in all countries except Japan fell within a narrow range based on average time spent per independent problem. In Japan, the average time per independent problem in most lessons was between approximately 10 and 20 minutes, but lessons with average problem times as long as 30 minutes were not uncommon. In the other countries, most lessons had average independent problem times between approximately 1 and 5 minutes and lessons with average problem times longer than approximately 10 minutes were uncommon.

Another way to examine the time spent on problems is to ask what percentage of problems was worked through relatively quickly. Because mathematics problem was defined to include simple, even routine, exercises, it could be the case that some problems, even a substantial percentage of problems, were worked through quickly. One would not necessarily expect these kinds of problems to provide the same learning opportunities as those that, for whatever reason, required more time to complete (National Research Council 2001a).

Problems that were worked out relatively quickly (less than 45 seconds) were distinguished from those that engaged students for longer periods of time (more than 45 seconds). The length of 45 seconds represented the consensus judgment of the mathematics code development team regarding a criterion that might separate many of the more routine exercises in the sample of eighth-grade lessons from those that involved more extensive work. Included in this analysis were all problems except for answered-only problems and concurrent problems for which no solution was presented publicly. For the concurrent problems in which a solution was publicly presented, the amount of time spent publicly discussing the problem could be computed (and determined to be greater or less than 45 seconds). Figure 3.7 presents the percentage of independent and concurrent problems that exceeded 45 seconds per eighth-grade mathematics lesson in each country.

FIGURE 3.7. Average percentage of independent and concurrent problems per eighth-grade mathematics lesson that were worked on for longer than 45 seconds, by country: 1999

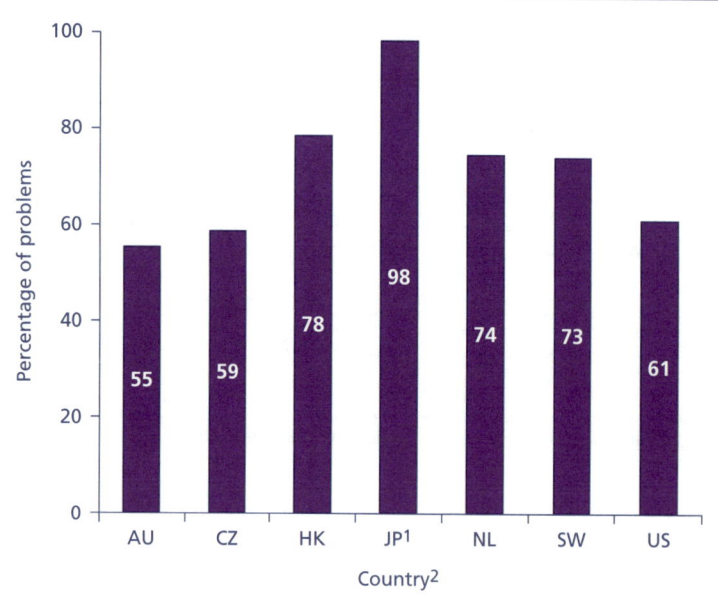

Country	Percentage
AU	55
CZ	59
HK	78
JP[1]	98
NL	74
SW	73
US	61

[1] Japanese mathematics data were collected in 1995.
[2] AU=Australia; CZ=Czech Republic; HK=Hong Kong SAR; JP=Japan; NL=Netherlands; SW=Switzerland; and US=United States.
NOTE: HK>AU, CZ, US; JP>AU, CZ, HK, NL, SW, US; NL>AU; SW>AU, CZ. Concurrent problems with no publicly presented solution were excluded. For each country, average percentage was calculated as the sum of the percentage within each lesson, divided by the number of lessons.
SOURCE: U.S. Department of Education, National Center for Education Statistics, Third International Mathematics and Science Study (TIMSS), Video Study, 1999.

In all the countries, the majority of problems per lesson for which time could be reliably determined extended beyond 45 seconds. Almost all of the problems in Japan met this threshold criterion (98 percent), a higher percentage than any other country. As stated earlier, on average, Japan also had the least number of independent problems worked on in each lesson and the longest time spent on each independent problem. Australia had a significantly lower percentage of problems per lesson that lasted at least 45 seconds than Hong Kong SAR, Japan, the Netherlands, and Switzerland.

By itself, time spent on problems says relatively little about the learning experiences of students. But like other indicators in this chapter, it provides a kind of parameter that can enable and constrain students' experiences. It might be difficult, for example, for students to solve a challenging problem, to examine the details of mathematical relationships that are revealed in the problem, or to discuss with the teacher and peers the reasons that solution methods work as they do if the problem is completed quickly (National Research Council 2001a).

The Purpose of Different Lesson Segments

Mathematical problems, together with non-problem segments, can be used by teachers to accomplish different purposes. And different countries might define these purposes in somewhat different ways. In consultation with the National Research Coordinators in each participating country, the following three purposes were defined:

- *Reviewing:* This category, more technically called "addressing content introduced in previous lessons," focused on the review or reinforcement of content presented previously [Video clip example 3.7]. These segments typically involved the practice or application of a topic learned in a prior lesson, or the review of an idea or procedure learned previously. Examples included:

    ○ Warm-up problems and games, often presented at the beginning of a lesson;

    ○ Review problems intended to prepare students for the new content;

    ○ Teacher lectures to remind students of previously learned content;

    ○ Checking the answers for previously completed homework problems; and

    ○ Quizzes and grading exercises.

- *Introducing new content:* This category focused on introducing content that students had not worked on in an earlier lesson [Video clip example 3.8]. Examples of segments of this type included:

    ○ Teacher expositions, demonstrations, and illustrations;

    ○ Teacher and student explorations through solving problems that were different, at least in part, from problems students had worked previously;

    ○ Class discussions of new content; and

    ○ Reading textbooks and working through new problems privately.

- *Practicing new content:* This category focused on practicing or applying content introduced in the current lesson [Video clip example 3.9]. These segments only occurred in lessons where new content was introduced. They typically took one of two forms: the practice or application of a topic already introduced in the lesson, or the follow-up discussion of an idea or formula after the class engaged in some practice or application. Examples of segments included:

- Working on problems to practice or apply ideas or procedures introduced in an earlier lesson;
- Class discussions of problem methods and solutions previously presented; and
- Teacher lectures summarizing or drawing conclusions about the new content presented earlier.

Segments coded as non-mathematical activity or mathematical organization were incorporated into the immediately following purpose segment, except when they appeared at the end of a lesson in which case they were included in the immediately preceding purpose segment. In this manner, all events in a lesson were classified as one (and only one) of the three purpose types. Only if the purpose of a segment was not clear was it coded "unable to make a judgment."

The entire eighth-grade mathematics lesson might have had the same purpose throughout, or it might have been segmented into different purposes. Figure 3.8 displays the average percentage of lesson time devoted to each of the three purpose types.

FIGURE 3.8. Average percentage of eighth-grade mathematics lesson time devoted to various purposes, by country: 1999

[1]Japanese mathematics data were collected in 1995.
[2]AU=Australia; CZ=Czech Republic; HK=Hong Kong SAR; JP=Japan; NL=Netherlands; SW=Switzerland; and US=United States.
[3]Practicing new content: HK>CZ, JP, SW.
[4]Introducing new content: HK, SW>CZ, US; JP>AU, CZ, HK, NL, SW, US.
[5]Reviewing: CZ>AU, HK, JP, NL, SW; US>HK, JP.
NOTE: For each country, average percentage was calculated as the sum of the percentage within each lesson, divided by the number of lessons. Percentages may not sum to 100 because of rounding and the possibility of coding portions of lessons as "not able to make a judgment about the purpose."
SOURCE: U.S. Department of Education, National Center for Education Statistics, Third International Mathematics and Science Study (TIMSS), Video Study, 1999.

On average, a higher percentage of lesson time in the Czech Republic was devoted to reviewing compared to all the other countries except the United States. More of the lesson time in Japan was spent introducing new content relative to all the other countries. And more of the lesson time in Hong Kong SAR was spent practicing new content than in the Czech Republic, Japan, and Switzerland.

By combining the time spent on the two purposes dealing with new content (introducing and practicing), it is possible to compare the time spent working on new content with the time spent reviewing content introduced in a prior lesson. This comparison can be made visually by comparing the lower section of figure 3.8 with the sum of the two upper sections. On average, in Australia, Hong Kong SAR, Japan, the Netherlands, and Switzerland a greater percentage of eighth-grade mathematics lesson time was spent on new material relative to previously learned material. In the Czech Republic, the reverse occurred. In the United States there was no difference found between the average amount of time spent reviewing older material and working on new material.

Another way to uncover the relative emphasis placed on reviewing previously presented content, introducing new content, and practicing new content is to check the percentage of lessons in which an activity of each type appeared. Did some lessons, for example, contain no review, or no new content?

As seen in table 3.4, 100 percent of the lessons in the Czech Republic sample included at least one review segment. The percentage was higher than all the countries except the United States. A higher percentage of Japanese lessons contained at least one segment introducing new content than all the other countries except Hong Kong SAR and Switzerland. And more Hong Kong SAR lessons included at least one segment devoted to practicing new content than Japan, the Netherlands, and Switzerland. These data appear to reinforce the pattern suggested in figure 3.4: Compared to some of the other countries, eighth-grade mathematics teachers in the Czech Republic (and to a lesser extent, in the United States) placed a greater emphasis on reviewing previously learned content; teachers in Japan placed a greater emphasis on introducing new content; and teachers in Hong Kong SAR placed a greater emphasis on practicing new content.

TABLE 3.4. Percentage of eighth-grade mathematics lessons with at least one segment of each purpose type, by country: 1999

Country	Reviewing[2]	Introducing new content[3]	Practicing new content[4]
Australia (AU)	89	71	58
Czech Republic (CZ)	100	75	63
Hong Kong SAR (HK)	82	92	81
Japan[1] (JP)	73	95	43
Netherlands (NL)	65	71	46
Switzerland (SW)	75	81	55
United States (US)	94	72	57

[1] Japanese mathematics data were collected in 1995.
[2] Reviewing: AU, US>NL, SW; CZ>AU, HK, JP, NL, SW.
[3] Introducing new content: JP>AU, CZ, NL, US; HK>CZ, US.
[4] Practicing new content: HK>JP, NL, SW.
SOURCE: U.S. Department of Education, National Center for Education Statistics, Third International Mathematics and Science Study (TIMSS), Video Study, 1999.

An additional lens through which to view the distribution of lesson segments devoted to review of old content versus the introduction and practice of new content is the percentage of lessons that focused on only one purpose. A question of special interest is whether any eighth-grade mathematics lessons were limited only to review. Such lessons would seem more likely to provide students with opportunities to become more familiar and efficient with content they have already encountered and less likely to have opportunities to learn new material.

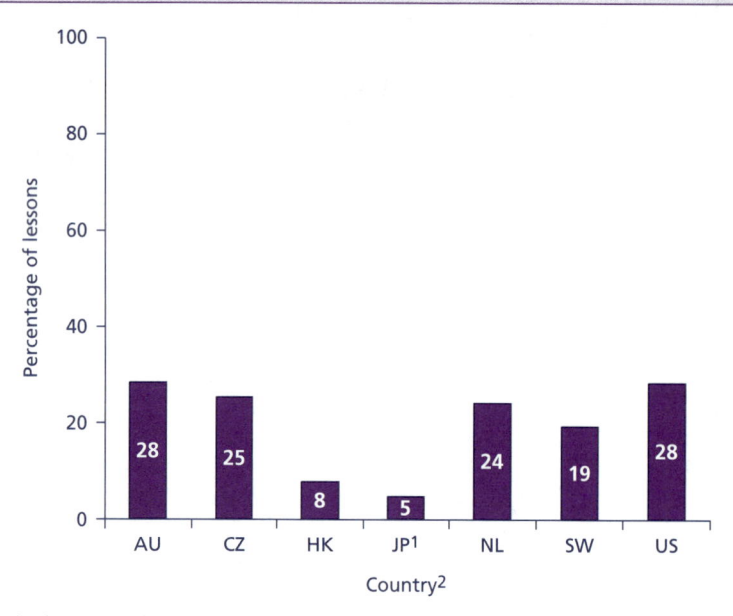

FIGURE 3.9. Percentage of eighth-grade mathematics lessons that were entirely review, by country: 1999

Country	Percentage
AU	28
CZ	25
HK	8
JP[1]	5
NL	24
SW	19
US	28

[1] Japanese mathematics data were collected in 1995.
[2] AU=Australia, CZ= Czech Republic, HK=Hong Kong SAR, JP=Japan, NL= Netherlands, SW=Switzerland, and US=United States.
NOTE: CZ, US>HK, JP. The tests for significance take into account the standard error for the reported differences. Thus, a difference between averages of two countries may be significant while the same difference between two other countries may not be significant.
SOURCE: U.S. Department of Education, National Center for Education Statistics, Third International Mathematics and Science Study (TIMSS), Video Study, 1999.

The relative emphasis on review in eighth-grade mathematics lessons in the Czech Republic and, to a lesser extent the United States, suggested in figure 3.8 and table 3.4 is reinforced in figure 3.9: Eighth-grade mathematics teachers in the Czech Republic and the United States devoted more lessons entirely to review than teachers in Hong Kong SAR and Japan.

In contrast to spending an entire lesson on one purpose, teachers might divide the lesson into segments that each focus on a different purpose. The number, length, and sequence of purpose segments provide one way of organizing lessons for accomplishing different goals. One measure of this lesson organization is simply the number of purpose segments. Different learning experiences might be afforded by lessons that have relatively short purpose segments and jump back and forth between segments of different purposes than by lessons that spend a longer time on a particular purpose before shifting to another. Table 3.5 shows the average number of shifts in purpose per lesson for each country.

TABLE 3.5. Average number of shifts in purpose per eighth-grade mathematics lesson, by country: 1999

Country	Average number of shifts in purpose[2]
Australia (AU)	2
Czech Republic (CZ)	2
Hong Kong SAR (HK)	3
Japan[1] (JP)	1
Netherlands (NL)	1
Switzerland (SW)	2
United States (US)	2

[1]Japanese mathematics data were collected in 1995.
[2]AU, CZ, SW, US>NL; HK>CZ, JP, NL, SW; AU, CZ>JP.
NOTE: The tests for significance take into account the standard error for the reported differences. Thus, a difference between averages of two countries may be significant while the same difference between two other countries may not be significant.
SOURCE: U.S. Department of Education, National Center for Education Statistics, Third International Mathematics and Science Study (TIMSS), Video Study, 1999.

The data in table 3.5 show that, on average, lessons in all the countries contained activities of at least two different purpose types (one shift equals moving from one purpose type to another). The results also indicate that, on average, eighth-grade mathematics teachers in Hong Kong SAR shifted among purposes more than teachers in the Czech Republic, Japan, the Netherlands, and Switzerland; in contrast, teachers in the Netherlands shifted among purposes less often than teachers in all the other countries except Japan.

It is possible to deduce from table 3.4 and figure 3.9 that the majority of lessons in all the countries contained at least two different kinds of purpose segments. The data in table 3.5 indicate that teachers rarely jumped between these purpose segments multiple times. Rather, it appears that, on average, eighth-grade mathematics teachers in each country often treated one purpose at length before shifting to another purpose.

Public and Private Classroom Interaction

Another element of the classroom environment that can enable and constrain different kinds of learning experiences for students is the way in which the teacher and students interact (Brophy 1999). Many classrooms include both whole-class discussions or public work, in which the teacher and students interact publicly, with the intent that all students participate (at least by listening), and private work, in which students complete assignments individually, or in small groups, and during which the teacher often circulates around the room and assists students who need help.

After viewing a number of the eighth-grade mathematics lessons in the TIMSS 1999 Video Study sample, the mathematics code development team observed that some teachers in the seven participating countries occasionally used interaction types different from these two. To capture all the interaction structures, five types of classroom interaction were defined:

- *Public interaction:* Public presentation by the teacher or one or more students intended for all students [Video clip example 3.10].

- *Private interaction:* All students work at their seats, either individually, in pairs, or in small groups, while the teacher often circulates around the room and interacts privately with individual students [Video clip example 3.11].

- *Optional, student presents information:* A student presents information publicly in written form, sometimes accompanied by verbal interaction between the student and the teacher or other students about the written work; other students may attend to this information or work on an assignment privately [Video clip example 3.12].

- *Optional, teacher presents information:* The teacher presents information publicly, in either verbal or written form, and students may attend to this information or work on an assignment privately.

- *Mixed private and public work:* The teacher divides the class into groups—some students are assigned to work privately on problems, while others work publicly with the teacher.

These interaction types were mutually exclusive and exhaustive; each segment of lesson time was classified as a single type. Table 3.6 displays the average percentage of lesson time devoted to public, private, and "optional, student presents information."[4] "Optional, teacher presents information" and "mixed private and public work" together accounted for no more than 2 percent of the lesson time in each country, on average, and are not shown in the table.

TABLE 3.6. Average percentage of eighth-grade mathematics lesson time devoted to public interaction, private interaction, and optional, student presents information, by country: 1999

Country	Public interaction[2]	Private interaction[3]	Optional, student presents information[4]
		Percent	
Australia (AU)	52	48	#
Czech Republic (CZ)	61	21	18
Hong Kong SAR (HK)	75	20	5
Japan[1] (JP)	63	34	3
Netherlands (NL)	44	55	‡
Switzerland (SW)	54	44	1
United States (US)	67	32	1

#Rounds to zero.
‡Reporting standards not met. Too few cases to be reported.
[1]Japanese mathematics data were collected in 1995.
[2]Public interaction: CZ>NL; HK>AU, CZ, JP, NL, SW; JP>AU, NL; US>AU, NL, SW.
[3]Private interaction: AU, SW>CZ, HK, JP, US; JP, US>CZ, HK; NL>CZ, HK, JP, SW, US.
[4]Optional, student presents information: CZ>AU, HK, JP, SW, US; JP>AU; HK>AU, SW, US.
NOTE: For each country, average percentage was calculated as the sum of the percentage within each lesson, divided by the number of lessons.
SOURCE: U.S. Department of Education, National Center for Education Statistics, Third International Mathematics and Science Study (TIMSS), Video Study, 1999.

Although in all the countries the vast majority of class time was spent in either public or private interaction, countries divided their time between them somewhat differently. Comparing across countries, eighth-grade mathematics lessons in Hong Kong SAR spent a greater percentage of

---

[4]As a consequence of taping procedures described in the technical report (Jacobs et al. forthcoming) and standards for obtaining cross-national reliability, more codes were developed for the public than the private portions of the lessons.

lesson time in public interaction (75 percent) than those in the other countries, except the United States. In the Netherlands, a greater percentage of lesson time (55 percent) was spent in private interaction compared to lesson time in the other countries, except Australia. When looking at how lesson time was divided among these interaction categories in each country, a greater percentage of mathematics lesson time was spent in public interaction than in either private interaction or in the mixed type of interaction referred to as optional, student presents information, except in the case of the Netherlands (as noted above) and Australia. In the case of Australia, there was no detectable difference between the percentage of lesson time spent in public and private interaction.

The Czech Republic was the only country to spend a substantial portion of time (18 percent) in the mixed type referred to as optional, student presents information. A prominent feature of the lessons in the Czech Republic was the public grading of students at the beginning of lessons. One or two students might have been called upon to publicly exhibit mastery of knowledge and skills taught previously, while the rest of their classmates worked privately. It is likely that the relatively long time Czech mathematics classes spent in this interaction type was related to this grading process.

As noted earlier, private interaction was defined as the time when students were working individually, in pairs, or in small groups. How often did students work alone? How often did they work with their peers? Figure 3.10 displays the average percentage of private interaction time during which students worked individually, or in pairs and groups.

FIGURE 3.10. Average percentage of private interaction time per eighth-grade mathematics lesson that students spent working individually or in pairs and groups, by country: 1999

Country	Worked in pairs and groups[3]	Worked individually[4]
AU	27	73
CZ	8	92
HK	5	95
JP[1]	24	76
NL	11	90
SW	26	74
US	20	80

[1] Japanese mathematics data were collected in 1995.
[2] AU=Australia; CZ=Czech Republic; HK=Hong Kong SAR; JP=Japan; NL=Netherlands; SW=Switzerland; and US=United States.
[3] Worked in pairs and groups: AU, JP, SW>CZ, HK, US>HK.
[4] Worked individually: CZ>AU, JP, SW; HK>AU, JP, SW, US; NL>SW.

NOTE: For each country, average percentage was calculated as the sum of the percentage within each lesson, divided by the number of lessons. The tests for significance take into account the standard error for the reported differences. Thus, a difference between averages of two countries may be significant while the same difference between two other countries may not be significant.

SOURCE: U.S. Department of Education, National Center for Education Statistics, Third International Mathematics and Science Study (TIMSS), Video Study, 1999.

Across all the countries, on average, at least 73 percent of private work time involved students completing tasks individually. The percentages ranged from 73 percent in Australia to 95 percent in Hong Kong SAR. Comparing percentages of time within countries shows that working individually is a more common activity for students in all the countries than working together during private work time.

Earlier it was noted that one way to examine the organization of eighth-grade mathematics lessons is to look at the number of purpose segments they contain. Similarly, varying the type of interaction provides another way for the teacher to structure the lesson and to emphasize different kinds of experiences. By shifting between interaction types, the teacher can modify the environment and ask students to work on mathematics in different ways. How often do eighth-grade mathematics teachers actually change interaction types (i.e., switch among the five categories listed earlier) during a lesson? Table 3.7 provides the relevant data.

TABLE 3.7. Average number of classroom interaction shifts per eighth-grade mathematics lesson, by country: 1999

Country	Average number of shifts[2]
Australia (AU)	5
Czech Republic (CZ)	7
Hong Kong SAR (HK)	5
Japan[1] (JP)	8
Netherlands (NL)	3
Switzerland (SW)	5
United States (US)	5

[1]Japanese mathematics data were collected in 1995.
[2]AU, HK, SW, US>NL; CZ, JP>AU, HK, NL, SW, US.
SOURCE: U.S. Department of Education, National Center for Education Statistics, Third International Mathematics and Science Study (TIMSS), Video Study, 1999.

Eighth-grade mathematics teachers in Japan and the Czech Republic made more shifts in interaction types than teachers in all the other countries. On average, they changed interaction types between 7 and 8 times each lesson. Using the median length of a mathematics lesson in these two countries (see table 3.1), each interaction segment lasted, on average, between 6 and 7 minutes. The fewest shifts occurred in the Netherlands, where interaction types changed an average of 3 times each lesson. On average, each interaction segment in the Netherlands lessons lasted 14 minutes.

For all the countries, the number of interaction shifts was significantly greater than the number of purpose shifts. It appears that teachers in all the countries used changes in interaction types to vary the learning environment more often than changes in the purpose of the activity.

The Role of Homework

The decision to incorporate homework within a lesson can directly impact how that lesson is organized. That is, teachers can review problems students completed prior to the lesson, allow students to begin homework problems assigned for a future lesson, or both. This section examines how frequently teachers assigned homework, and the extent to which homework was worked on as part of the lesson activities.

Figure 3.11 displays for each country the percentage of eighth-grade mathematics lessons in which homework was assigned.

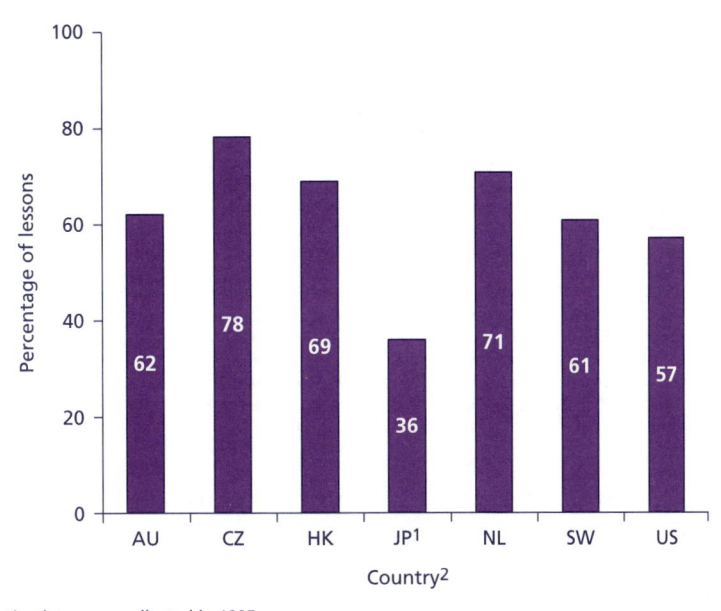

FIGURE 3.11. Percentage of eighth-grade mathematics lessons in which homework was assigned, by country: 1999

[1]Japanese mathematics data were collected in 1995.
[2]AU=Australia; CZ=Czech Republic; HK=Hong Kong SAR; JP=Japan; NL=Netherlands; SW=Switzerland; and US=United States.
NOTE: AU, CZ, HK, NL, SW>JP.
SOURCE: U.S. Department of Education, National Center for Education Statistics, Third International Mathematics and Science Study (TIMSS), Video Study, 1999.

Across the countries, except Japan, homework was assigned in at least 57 percent of the lessons. Japanese eighth-grade mathematics teachers assigned homework less often (in 36 percent of lessons) than teachers in all the other countries except the United States.

More information can be provided about the role of homework by examining whether students were allowed to begin the assignment during the lesson [Video clip example 3.13]. The first column in table 3.8 displays the average number of problems per eighth-grade mathematics lesson assigned as homework that were begun in the lesson. The second column displays an estimate of the average time spent on these problems per lesson.

TABLE 3.8.	Average number of eighth-grade mathematics problems per lesson assigned as homework and begun in the lesson, and estimated average time per lesson spent on these problems, by country: 1999

Country	Average number of problems per lesson assigned for homework[2]	Estimated average time per lesson spent on future homework (in minutes)[3,4]
Australia (AU)	5	4
Czech Republic (CZ)	1	2
Hong Kong SAR (HK)	2	3
Japan[1] (JP)	#	1
Netherlands (NL)	10	10
Switzerland (SW)	2	4
United States (US)	4	3

#Rounds to zero.
[1]Japanese mathematics data were collected in 1995.
[2]Average number of problems per lesson assigned for homework: AU, US>CZ, JP; HK, SW>JP; NL>CZ, HK, JP, SW, US.
[3]This number includes the exact amount of time spent on all answered-only and independent problems assigned as future homework, plus an estimate of the length of time spent on all concurrent problems assigned as future homework. This latter estimate was calculated for each lesson by dividing the total length of time spent on concurrent problems by the number of concurrent problems, and multiplying the result by the number of concurrent problems assigned as future homework.
[4]Estimated average time per lesson spent on future homework: AU>JP; NL>AU, CZ, HK, JP, SW, US.
NOTE: The tests for significance take into account the standard error for the reported differences. Thus, a difference between averages of two countries may be significant while the same difference between two other countries may not be significant.
SOURCE: U.S. Department of Education, National Center for Education Statistics, Third International Mathematics and Science Study (TIMSS), Video Study, 1999.

In eighth-grade Dutch lessons, students began work during lessons on a greater number of problems assigned as homework (10) on average than their peers in all the other countries except Australia. Moreover, in Dutch lessons, significantly more estimated lesson time (10 minutes) was spent on these problems than in all the other countries. The estimated average amount of time students spent beginning their homework assignment in all the other countries ranged from 1 minute in Japan to 4 minutes in Australia and Switzerland. Using median lesson times (from table 3.1), it appears that in all the countries, except the Netherlands, starting on the homework assignment filled, on average, no more than 10 percent of lesson time.

Another measure of the role that homework played in the lessons is the way in which problems completed for the videotaped lesson were corrected and discussed [Video clip example 3.14]. Table 3.9 displays the average number of problems per eighth-grade mathematics lesson that were previously assigned as homework and corrected or discussed during the videotaped lesson, along with an estimate of the average time spent on these problems.

TABLE 3.9. Average number of eighth-grade mathematics problems per lesson previously assigned as homework and corrected or discussed during the lesson, and estimated average time per lesson spent on these problems, by country: 1999

Country	Average number of problems per lesson previously assigned as homework[2]	Estimated average time per lesson spent on previously assigned homework (in minutes)[3,4]
Australia (AU)	3	1
Czech Republic (CZ)	#	#
Hong Kong SAR (HK)	#	1
Japan[1] (JP)	#	1
Netherlands (NL)	12	16
Switzerland (SW)	5	5
United States (US)	8	7

#Rounds to zero.
[1]Japanese mathematics data were collected in 1995.
[2]Average number of problems per lesson previously assigned as homework: NL>AU, CZ, HK, JP; US>CZ, HK, JP.
[3]This number includes the exact amount of time spent on all answered-only and independent problems previously assigned as homework, plus an estimate of the length of time spent on all concurrent problems previously assigned as homework. This latter estimate was calculated for each lesson by dividing the total length of time spent on concurrent problems by the number of concurrent problems, and multiplying the result by the number of concurrent problems previously assigned as homework.
[4]Estimated average time per lesson spent on previously assigned homework: NL>AU, CZ, HK, JP.
SOURCE: U.S. Department of Education, National Center for Education Statistics, Third International Mathematics and Science Study (TIMSS), Video Study, 1999.

Teachers and students in the Netherlands corrected and discussed more problems per lesson that were previously assigned as homework (12) on average and spent more estimated time on these problems (16 minutes) than students and teachers in all the other countries except Switzerland and the United States.

Considering both work on past and future homework problems, homework was treated as a more central part of the eighth-grade mathematics lessons in the Netherlands than in most other countries. Teachers in Australia, Switzerland, and the United States showed some indication of attending to homework during the lesson, but homework was a relatively minor part of the lesson, on average, in the Czech Republic, Hong Kong SAR, and Japan.

Pedagogical Features That Influence Lesson Clarity and Flow

A final set of structural elements of the lesson environment considered in this chapter focuses on lesson flow and clarity. These include lesson features that seem to highlight the major points of the lesson for the students or, on the other hand, might interrupt the flow of the lesson.

Goal Statements

One way that teachers can help students identify the key mathematical points of a lesson is to describe the goal of the lesson (Brophy 1999). Goal statements were defined as explicit written or verbal statements by the teacher about the specific mathematical topic(s) that would be covered during the lesson [Video clip example 3.15]. To count as a goal statement, the statement had to preview the mathematics that students encountered during at least one-third of the lesson time. Figure 3.12 displays the percentage of eighth-grade mathematics lessons in each country that contained at least one goal statement.

FIGURE 3.12. Percentage of eighth-grade mathematics lessons that contained at least one goal statement, by country: 1999

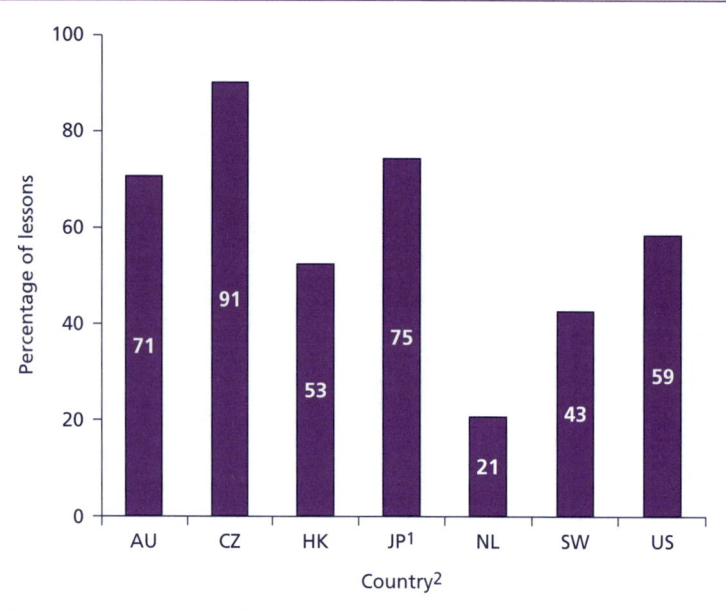

[Bar chart showing percentage of lessons by country: AU=71, CZ=91, HK=53, JP=75, NL=21, SW=43, US=59]

[1]Japanese mathematics data were collected in 1995.
[2]AU=Australia; CZ=Czech Republic; HK=Hong Kong SAR; JP=Japan; NL=Netherlands; SW=Switzerland; and US=United States.
NOTE: AU>NL, SW; CZ>AU, HK, NL, SW, US; HK, JP, SW, US>NL.
SOURCE: U.S. Department of Education, National Center for Education Statistics, Third International Mathematics and Science Study (TIMSS), Video Study, 1999.

A higher percentage of eighth-grade mathematics lessons in the Czech Republic contained goal statements provided by the teacher (91 percent) than in all the other countries except Japan. In the Netherlands, goal statements were provided in a lower percentage of lessons (21 percent) than in all the other countries.

Lesson Summary Statements

A second kind of aid to help students recognize the key ideas in a lesson is a summary statement. Summary statements highlight points that have just been studied in the lesson. They were defined as statements that occurred near the end of the public portions of the lesson and described the mathematical point(s) of the lesson [Video clip example 3.16]. Figure 3.13 displays the percentage of eighth-grade mathematics lessons in each country that contained at least one summary statement.

FIGURE 3.13. Percentage of eighth-grade mathematics lessons that contained at least one summary statement, by country: 1999

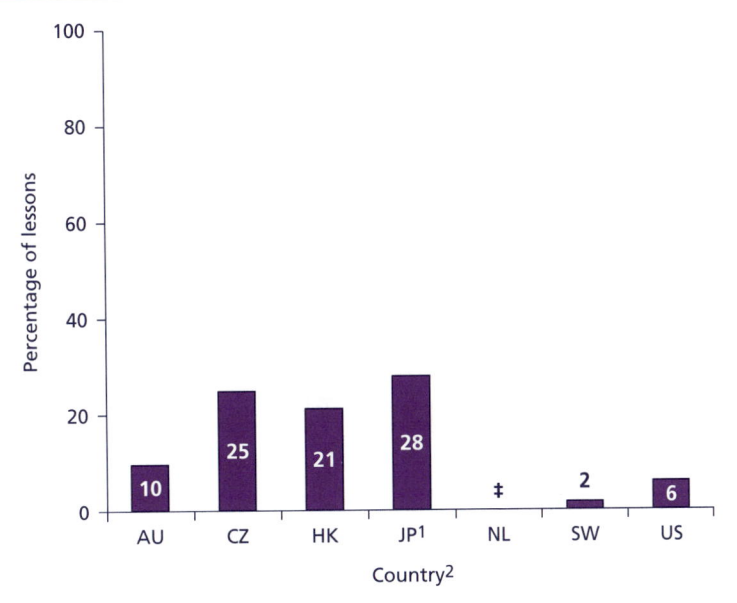

Country	Percentage
AU	10
CZ	25
HK	21
JP[1]	28
NL	‡
SW	2
US	6

‡Reporting standards not met. Too few cases to be reported.
[1]Japanese mathematics data were collected in 1995.
[2]AU=Australia; CZ=Czech Republic; HK=Hong Kong SAR; JP=Japan; NL=Netherlands; SW=Switzerland; and US=United States.
NOTE: CZ>SW, US; HK, JP>SW. The tests for significance take into account the standard error for the reported differences. Thus, a difference between averages of two countries may be significant while the same difference between two other countries may not be significant.
SOURCE: U.S. Department of Education, National Center for Education Statistics, Third International Mathematics and Science Study (TIMSS), Video Study, 1999.

For all the countries, lesson summaries were less common than goal statements. Lesson summaries were found in at least 21 percent of eighth-grade mathematics lessons in Japan, the Czech Republic, and Hong Kong SAR, and in 10 percent of lessons in Australia. In the other countries where reliable estimates could be calculated, between 2 and 6 percent of lessons included summary statements.

Outside Interruptions

Whereas goal statements and summary statements can enhance the clarity of the key lesson ideas, interruptions to the lesson can break its flow and, perhaps, interfere with or delay developing the key ideas (Stigler and Hiebert 1999). One kind of interruption comes from outside the classroom. Examples of outside interruptions include announcements over the intercom, individuals from outside the class requiring the teacher's attention, and talking to a student who has arrived late [Video clip example 3.17]. Figure 3.14 displays the percentage of eighth-grade mathematics lessons in which at least one outside interruption occurred.

FIGURE 3.14. Percentage of eighth-grade mathematics lessons with at least one outside interruption, by country: 1999

Country	Percentage
AU	30
CZ	14
HK	28
JP[1]	8
NL	32
SW	11
US	29

[1] Japanese mathematics data were collected in 1995.
[2] AU=Australia; CZ=Czech Republic; HK=Hong Kong SAR; JP=Japan; NL=Netherlands; SW=Switzerland; and US=United States.
NOTE: NL>JP. Percentage reported for Japan differs from that reported in Stigler et al. (1999) because the definition was changed for the current study.
SOURCE: U.S. Department of Education, National Center for Education Statistics, Third International Mathematics and Science Study (TIMSS), Video Study, 1999.

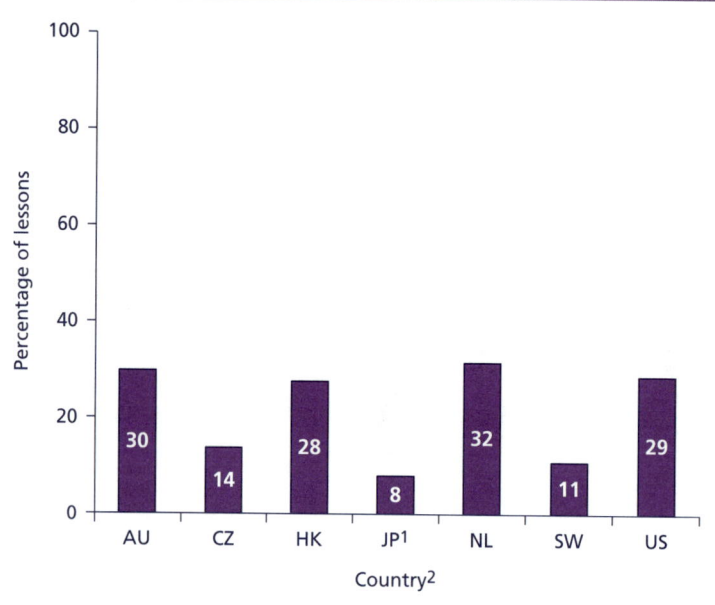

In all the countries some lessons were interrupted. In fact, about 30 percent of lessons were interrupted in Australia, Hong Kong SAR, the Netherlands, and the United States. A larger percentage of lessons were interrupted in the Netherlands than in Japan. Other apparent differences between countries were not significant.

Non-Mathematical Segments

Another type of potential interruption to the flow of lessons occurred when the class engaged in non-mathematical activities after the mathematics portion of the lesson had begun. By definition, a non-mathematical segment was any activity unrelated to the teaching and learning of mathematics that lasted for at least 30 seconds. When these segments occurred at the beginning of the lesson (e.g., when teachers and students attended to such things as greetings, checking attendance, or discussing the videotaping procedure) or at the end of lesson (e.g., when teachers and students talked socially, or discussed when and where future lessons would be held), they did not interrupt the flow of the lesson. So, the segments of interest here were those that occurred within the mathematics portion of the lesson.

An important characteristic of these interruptions was that they appeared, in some cases, to be within the teacher's control. As such, they differed from outside interruptions, which appeared to be largely outside of the teacher's control. Figure 3.15 displays the percentage of lessons that contained at least one non-mathematical work segment within the mathematics portion of the lesson.

FIGURE 3.15. Percentage of eighth-grade mathematics lessons with at least one non-mathematical segment at least 30 seconds in length within the mathematics portion of the lesson, by country: 1999

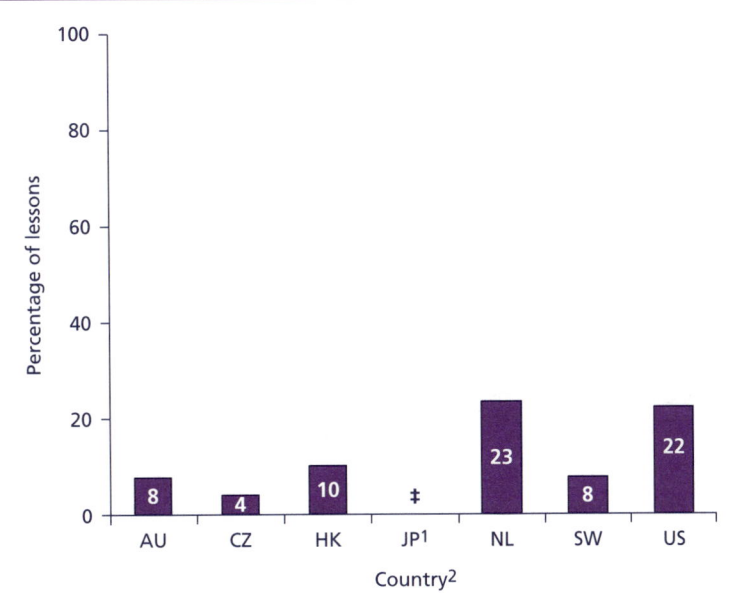

Country	Percentage
AU	8
CZ	4
HK	10
JP[1]	‡
NL	23
SW	8
US	22

‡Reporting standards not met. Too few cases to be reported.
[1]Japanese mathematics data were collected in 1995.
[2]AU=Australia; CZ=Czech Republic; HK=Hong Kong SAR; JP=Japan; NL=Netherlands; SW=Switzerland; and US=United States.
NOTE: NL, US>CZ; US>SW. The tests for significance take into account the standard error for the reported differences. Thus, a difference between averages of two countries may be significant while the same difference between two other countries may not be significant.
SOURCE: U.S. Department of Education, National Center for Education Statistics, Third International Mathematics and Science Study (TIMSS), Video Study, 1999.

The percentages of lessons containing these interruptions ranged from 23 percent and 22 percent of the eighth-grade mathematics lessons in the Netherlands and the United States, respectively, to 4 percent in the Czech Republic. There were too few lessons containing these interruptions in the Japanese lessons to calculate a reliable estimate. The percentage of lessons containing these interruptions in the Netherlands and the United States was higher than that in the Czech Republic. Additionally, the percentage in the United States was higher than that in Switzerland.

Public Announcements That Were Unrelated to the Current Mathematics Assignment

A second way in which teachers' actions might affect the flow of the lesson and potentially interrupt students' work occurred when teachers made an off-topic announcement during private work time. This type of announcement was defined as one containing either no mathematical information (for example, the teacher might have addressed a disciplinary problem) or mathematical information that appeared to be unrelated to the assignment at hand [Video clip example 3.18]. There was no minimum time length for this code, so a public announcement of this sort could be of varying length.

Figure 3.16 displays the percentage of eighth-grade mathematics lessons that contained at least one public announcement during private work time that appeared to be unrelated to the current assignment.

FIGURE 3.16. Percentage of eighth-grade mathematics lessons with at least one public announcement by the teacher during private work time unrelated to the current assignment, by country: 1999

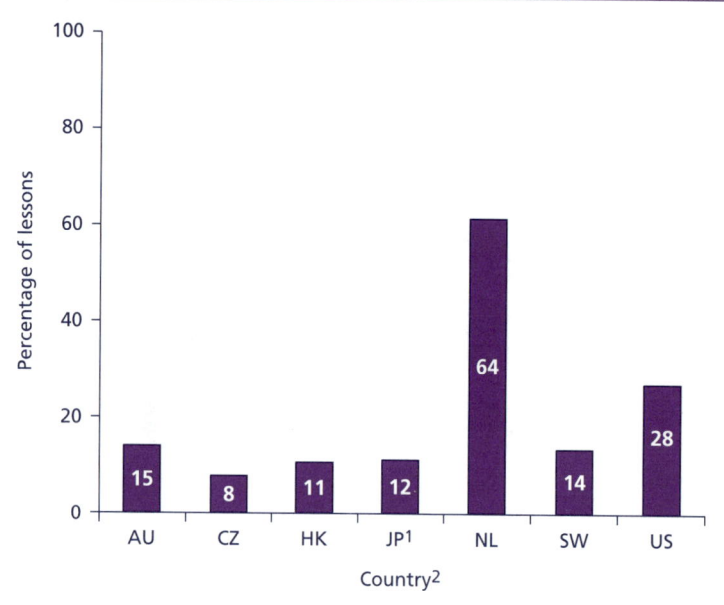

[Bar chart showing percentage of lessons by country: AU=15, CZ=8, HK=11, JP=12, NL=64, SW=14, US=28]

[1] Japanese mathematics data were collected in 1995.
[2] AU=Australia; CZ=Czech Republic; HK=Hong Kong SAR; JP=Japan; NL=Netherlands; SW=Switzerland; and US=United States.
NOTE: NL>AU, CZ, HK, JP, SW, US; US>CZ.
SOURCE: U.S. Department of Education, National Center for Education Statistics, Third International Mathematics and Science Study (TIMSS), Video Study, 1999.

In the Netherlands, more eighth-grade mathematics lessons contained public announcements unrelated to the current assignment (64 percent) than in all the other countries. Twenty-eight percent of U.S. lessons contained this kind of public announcement, a larger percentage than in the Czech Republic.

Summary

Teaching can be analyzed from many perspectives. The approach taken in this study was to focus on features of teaching, and the way these features fit together, that seem likely to influence the learning opportunities for students (Brophy 1999; National Research Council 1999, 2001a; Stigler et al. 1999). Four major categories of features were defined for this report: the context of the lesson, including information about the teacher and students; the organization and structure of the lesson environment; the kind of mathematics studied; and the way in which the mathematics was studied.

Chapter 2 presented information on the context of the lesson. In the current chapter, results were presented on elements of the lesson that structured the learning environment for students. As noted earlier, although these elements might not influence learning directly they might set boundaries on the kinds of learning experiences that were likely to occur. How the details were filled in matter a great deal, of course, and these details will be examined in chapters 4 and 5. The results of this chapter, nonetheless, represent some basic stage-setting choices that appeared in the eighth-grade mathematics lessons of each country. Consequently, these results provide a good beginning point for looking inside the classrooms of the seven participating countries.

At one level, it appears that educators in the seven countries made similar choices with respect to organizing lessons. They used many of the same basic ingredients. Virtually all eighth-grade lessons contained mathematical problems, and most of the instructional time was devoted to solving problems (figure 3.3). Some problems were presented for class discussion and some were assigned as a set for working on privately (table 3.3 and figure 3.4). Across all lessons, teachers devoted some time to reviewing old content, introducing new content, and practicing new content (figure 3.8). And work was accomplished through two primary social structures: working together as a whole class and working privately (table 3.6).

A closer look reveals, however, that there were detectable differences among countries in the relative emphasis they placed on different values of these variables. These differences are highlighted in the following summaries (after noting some important similarities). The reason for focusing on differences among countries is not because they necessarily represent better choices for teaching eighth-grade mathematics, but because they represent different choices. It is through examining different choices, perhaps choices not previously considered or even imagined, that educators can raise the level of discussion needed to make the improvement of teaching a systematic and collective professional activity.

Among the important findings reported in this chapter are the following:

- Eighth-grade mathematics teachers and students in every country spent a high percentage of lesson time engaged in mathematical work (figure 3.2).

- In all the participating countries, eighth-grade mathematics was taught predominantly through solving problems (figure 3.3). Apparently, this is the common currency of mathematics teachers in these countries. Some readers might regard this as an obvious finding, but the extent to which working on problems provides the prevalent instructional activity in countries around the world has not been documented previously.

- Japan and the Netherlands provided two comparatively distinct learning environments for students as defined by a few basic organizational features:

    ○ Japanese eighth-grade mathematics lessons focused on presenting new content through solving a few problems, mostly as a whole class, with each problem requiring a considerable length of time (tables 3.3 and 3.6, and figures 3.5 and 3.8);

    ○ In Dutch lessons, private work played a more central role, with eighth-grade students spending a larger percentage of time working on a set of problems, either reviewing old homework or starting on newly assigned homework (figure 3.4 and tables 3.6, 3.8, and 3.9).

    In these different structures, the teacher, the written curriculum materials, and the students would seem to play quite different roles.

- Countries emphasized different purposes in their eighth-grade mathematics lessons. Compared to some other countries, the Czech Republic (and to a lesser extent the United States) emphasized reviewing, whereas Hong Kong SAR and Japan emphasized new content, with Japan focusing on introducing the new content and Hong Kong SAR focusing on practicing the new content (figures 3.8 and 3.9, and table 3.4).

- Finally, the Czech Republic and the Netherlands showed different profiles with regard to lesson clarity and flow. Lessons in the Czech Republic were relatively high on measures that might aid students in identifying the key points of the lesson (e.g., goal statements and summary statements) and relatively low on the measures of potential interruption to lesson flow (e.g., outside interruptions, non-mathematical segments within the mathematics portion of the lesson, and unrelated public announcements). Lessons in the Netherlands showed the opposite profile (figures 3.12 through 3.16).

As noted several times, the results of this chapter are most important in the way they set the stage for the next two chapters. It will be worth keeping in mind the organizational elements prevalent in each country as the results of the next two chapters are studied. Many of those results are enabled and supported by the features of the lesson environments presented in this chapter.

CHAPTER 4
The Mathematical Content of Lessons

Students' opportunity to learn mathematics is shaped, in part, by the content of the mathematics presented (National Research Council 2001a). The structure of the classroom learning environments described in chapter 3 provides a shell that can enable and constrain particular kinds of learning opportunities. The nature of these opportunities is shaped further by the way in which the shell is filled. This chapter considers the content that filled the eighth-grade mathematics lessons and the next chapter focuses on the way in which the content was treated.[1]

At one level, the importance of content in shaping students' learning opportunities is obvious and rather easy to see. Students have little chance of learning algebra in school, for example, if algebra is not part of the lesson. This is why the analysis of curricula has played a major role in searching for relationships between classroom practice and student achievement in both the Second International Mathematics Study (McKnight et al. 1987) and the Third International Mathematics and Science Study (Schmidt et al. 1999).

The TIMSS 1999 Video Study provides an opportunity to examine the content of lessons in considerable detail. Moving beyond the intended curricula contained in syllabi and textbooks, the filmed lessons reveal the implemented curricula as well. The focus then becomes the mathematical content that students actually encountered in the classroom. Examination of the content can be conducted in a variety of ways, including analyses that provide detailed descriptions of the topics covered or that reveal the complexity of the topics as evidenced through the problems presented, the mathematical relationships among the problems, and the nature of the mathematical reasoning evident in the lesson.

As noted in chapter 1, it is inappropriate to draw conclusions that link directly the content of the filmed lessons and students' achievement. There are too many other factors that might affect the relative level of achievement in each country and, moreover, the sampling procedures prevent causal connections between the sample of filmed lessons and countrywide achievement. Rather, the findings in this chapter should be interpreted as choices that have been made about the nature of the content of eighth-grade mathematics lessons in each country. Uncovering these choices can help educators make more informed decisions about whether the choices currently being made in their country provide the desired learning opportunities for students.

---

[1] Additional descriptions of the mathematics content of a sub-sample of lessons from each country are presented in appendix D. The mathematics quality analysis group (see appendix A) examined lessons for indicators of mathematical quality, such as curricular level, the kind of reasoning required, mathematical coherence, and the extent to which mathematical ideas are developed. Because the group analyzed only a sub-sample of lessons, the results are considered experimental and are therefore presented in an appendix.

Mathematical Topics Covered During the Lessons

The filmed lessons in the TIMSS 1999 Video Study were obtained by sampling steadily over the school year (except for the 1995 Japanese sample).[2] This means it is reasonable to presume that each nation's sample is somewhat representative of the topics covered in these countries during eighth grade. But, because the sample was not chosen to represent systematically the curriculum in each country, it is not appropriate to treat this study as a test of curricular differences among countries. Consequently, the mathematical topics covered in each country's sample are described below but no statistical comparisons were made. Statistical comparisons were made, however, on other features of content such as the complexity of mathematical problems or relationships among problems in lessons, which are presented later in this chapter.

Because some lessons covered several mathematical topics, the most accurate way of describing the topics included in the lessons was to label each mathematical problem as pertaining to a specific topic. The mathematics problem analysis group constructed a detailed and comprehensive list of mathematics topics covered in eighth grade in all participating countries (see appendix A for a description of this coding group). Then, using written records of the lessons, each problem discussed individually during a lesson (independent problems) and each problem assigned to students as part of a group (concurrent problems) was labeled with an entry from this list. Nearly 15,000 mathematics problems were coded.

The topics addressed by the mathematics problems were grouped into five major categories and several sub-categories.

- *Number:* Whole numbers, fractions, decimals, ratio, proportion, percent, and integers;

- *Geometry:* Measurement (perimeter and area), two-dimensional geometry (polygons, angles, lines, transformations, and constructions), and three-dimensional geometry;

- *Statistics:* Probability, statistics, and graphical representation of data;

- *Algebra:* Operations with linear expressions, linear equations, inequalities and graphs of linear functions, and quadratic and higher degree equations; and

- *Trigonometry:* Trigonometric identities, equations with trigonometric expressions.

In some lessons, all of the problems were from one topic sub-category, such as linear equations, whereas in other lessons the problems were from more than one sub-category, and in some cases, more than one major category. Table 4.1 shows the average percentage per eighth-grade lesson of mathematical problems within each major content category and within sub-categories for number, geometry, and algebra. The percentages were calculated by averaging the percentages of problems of each type in each lesson across all lessons within a country. Thus, no single lesson is likely to show the distribution in table 4.1.

---

[2] As noted in chapter 1, most of the Japanese lessons were collected over a four-month period rather than over the full school year (Stigler et al. 1999).

Chapter 4
The Mathematical Content of Lessons

TABLE 4.1. Average percentage of problems per eighth-grade mathematics lesson within each major category and sub-category topic area, by country: 1999

Topic area	AU	CZ	HK	JP[2]	NL	SW	US
Number	36	27	18	‡	16	42	30
Whole numbers, fractions, decimals	15	13	5	‡	6	20	17
Ratio, proportion, percent	19	4	10	‡	6	19	6
Integers	2	9	3	‡	4	3	8
Geometry	29	26	24	84	32	33	22
Measurement (perimeter and area)	10	6	3	11	9	12	13
Two-dimensional geometry (polygons, angles, lines)	14	15	17	73	15	17	4
Three-dimensional geometry	5	6	5	‡	9	4	5
Statistics	9	3	2	‡	10	2	6
Algebra	22	43	40	12	41	22	41
Linear expressions	7	16	11	‡	6	5	6
Solutions and graphs of linear equations and inequalities	15	21	23	12	33	14	27
Higher-order functions	‡	6	6	‡	3	3	8
Trigonometry	‡	‡	14	‡	‡	‡	‡
Other	‡	1	‡	‡	‡	1	1

‡Reporting standards not met. Too few cases to be reported.
[1]AU=Australia; CZ=Czech Republic; HK=Hong Kong SAR; JP=Japan; NL=Netherlands; SW=Switzerland; and US=United States.
[2]Japanese mathematics data were collected in 1995.
NOTE: Percentages may not sum to 100 because of rounding and data not reported. For each country, average percentage was calculated as the sum of the percentage within each lesson, divided by the number of lessons.
SOURCE: U.S. Department of Education, National Center for Education Statistics, Third International Mathematics and Science Study (TIMSS), Video Study, 1999.

In all countries, at least 82 percent of the problems per eighth-grade mathematics lesson, on average, addressed three major areas—number, geometry, and algebra. In Japan, 73 percent of the problems per lesson, on average, dealt with two-dimensional geometry. This might be due to the fact that, as noted elsewhere, most of the lessons collected as part of the Japanese sample in the TIMSS 1995 Video Study were gathered over a portion of the school year (Stigler et al. 1999). Later analyses will examine the extent to which the high percentage of geometry problems was related to other content characteristics.

Level of Mathematics Evident in the Lessons

Mathematical topics within each major area often are ordered in the curriculum, with some topics following others. Understanding later topics often is facilitated by being familiar with earlier ones. For example, when learning algebra students usually study linear functions before quadratic functions. And students often learn to simplify linear expressions before they solve linear equations and inequalities. Within each of the three major areas—number, geometry, and algebra—it is possible to identify several levels of topics corresponding to the order in which they might be encountered in a curriculum. In table 4.1, the sub-categories for number and algebra, and the first two sub-categories for geometry, show one possible ordering. No levels of difficulty or complexity are suggested by the ordering of the major topic areas, except that trigonometry often is treated later in the curriculum than the other topics listed.

On average, 42 percent of problems per eighth-grade mathematics lesson in the Swiss sample dealt with number and about half of these (20 percent of problems per lesson) focused on the first-level sub-topics of whole numbers, fractions, and decimals. The lessons from Australia, the Czech Republic, and the United States included about 15 percent of problems per lesson that addressed whole numbers, fractions, and decimals.

In all countries' samples, except that of the United States, at least 14 percent of problems per eighth-grade mathematics lesson were at a more advanced level of two-dimensional geometry—polygons, angles, lines, transformations, and constructions—and from 3 to 12 percent of problems were at a less advanced level—measuring perimeters and two-dimensional areas. The United States lessons showed the reverse pattern, devoting about 4 percent of problems per lesson to a more advanced level and 13 percent of problems to a less advanced level. Japan's sample of lessons focused mostly on the more advanced two-dimensional geometry topics.

On average, about 40 percent of the mathematics problems per eighth-grade lesson in the samples of the Czech Republic, Hong Kong SAR, the Netherlands, and the United States dealt with algebra. In Australia, Switzerland, and Japan, 22 percent or less of the problems per lesson involved algebra. Across all countries, between 12 and 33 percent of problems per lesson focused on mid-level algebra topics—solving linear equations and inequalities and graphing linear functions—whereas between 3 and 8 percent of problems per lesson focused on higher-order functions, where reliable estimates could be calculated.

Trigonometry was involved in 14 percent of the problems per eighth-grade lesson, on average, in the Hong Kong SAR sample. In the mathematics lessons of the other countries, trigonometry problems occurred too infrequently to report reliable estimates.

Type of Mathematics Evident in the Lessons

Two characteristics of the mathematics presented during the lessons are its complexity and the kind of reasoning that is involved when doing the mathematics. Two aspects of these characteristics will be examined here and then considered further in appendix D.

Procedural Complexity

The complexity of the mathematics presented in the lessons is an important feature of the mathematics but it is difficult to define and code reliably. The complexity of a problem depends on a number of factors, including the experience and capability of the student. One kind of complexity that can be defined independent of the student is procedural complexity—the number of steps it takes to solve a problem using a common solution method.

The mathematics problem analysis group (see appendix A) developed a scheme for coding procedural complexity and analyzed every problem worked on or assigned during each eighth-grade mathematics lesson (independent and concurrent problems, see chapter 3). Problems were sorted into low, moderate, or high complexity according to the following definitions:

- *Low complexity:* Solving the problem, using conventional procedures, requires four or fewer decisions by the students (decisions could be considered small steps) [Video clip example 4.1]. The problem contains no sub-problems, or tasks embedded in larger problems that could themselves be coded as problems.

  ○ Example: Solve the equation: $2x + 7 = 2$.

- *Moderate complexity:* Solving the problem, using conventional procedures, requires more than four decisions by the students and can contain one sub-problem [Video clip example 4.2].

  ○ Example: Solve the set of equations for x and y: $2y = 3x - 4$; $2x + y = 5$.

- *High complexity:* Solving the problem, using conventional procedures, requires more than four decisions by the students and contains two or more sub-problems [Video clip example 4.3].

  ○ Example: Graph the following linear inequalities and find the area of intersection: $y \leq x + 4$; $x \leq 2$; $y \geq -1$.

Figure 4.1 shows the average percentage of problems per eighth-grade mathematics lesson that were of each complexity level.

FIGURE 4.1. Average percentage of eighth-grade mathematics problems per lesson at each level of procedural complexity, by country: 1999

Country	Low complexity	Moderate complexity	High complexity
AU	77	16	8
CZ	64	25	11
HK	63	29	8
JP[1]	17	45	39
NL	69	25	6
SW	65	22	12
US	67	27	6

[1] Japanese mathematics data were collected in 1995.
[2] AU=Australia; CZ=Czech Republic; HK=Hong Kong SAR; JP=Japan; NL=Netherlands; SW=Switzerland; and US=United States.
[3] High complexity: JP>AU, CZ, HK, NL, SW, US.
[4] Moderate complexity: HK>AU; JP>AU, SW.
[5] Low complexity: AU, CZ, HK, NL, SW, US>JP.

NOTE: Percentages may not sum to 100 because of rounding. For each country, average percentage was calculated as the sum of the percentage within each lesson, divided by the number of lessons.
SOURCE: U.S. Department of Education, National Center for Education Statistics, Third International Mathematics and Science Study (TIMSS), Video Study, 1999.

In each country, except Japan, at least 63 percent of the mathematics problems per lesson, on average, were of low procedural complexity and up to 12 percent of the problems were of high procedural complexity. Japanese eighth-grade mathematics lessons, in contrast, contained relatively fewer problems of low complexity and about equal percentages of moderate and high complexity (17 percent, 45 percent, and 39 percent, respectively). Differences between Japan and all the other countries were found at both the low and high levels of procedural complexity.

Because the Japanese sample contained lessons with high percentages of two-dimensional geometry problems relative to the other countries, a question is raised about whether the relatively high complexity profile in Japan was due to the topic sample. Figure 4.2 shows the average percentage of two-dimensional geometry problems per eighth-grade mathematics lesson of each complexity level in each country. These percentages are based only on lessons that contained two-dimensional geometry problems; therefore, the sample size was much smaller than the full sample. Because of the reduced sample sizes, the results should be interpreted with caution.

FIGURE 4.2. Average percentage of two-dimensional geometry problems at each level of procedural complexity per eighth-grade mathematics lesson in sub-sample of lessons containing two-dimensional geometry problems, by country: 1999

[1]Japanese mathematics data were collected in 1995.
[2]AU=Australia; CZ=Czech Republic; HK=Hong Kong SAR; JP=Japan; NL=Netherlands; SW=Switzerland; and US=United States.
[3]High complexity: JP>AU, HK.
[4]Moderate complexity: No differences detected.
[5]Low complexity: AU, HK, NL>JP.
NOTE: Because the focus of this analysis was on two-dimensional geometry problems, only those lessons that contained such problems were included. Thus, the sample sizes for each country are less than the full sample. Results should therefore be interpreted with caution. For each country, average percentage was calculated as the sum of the percentage within each lesson, divided by the number of lessons. Percentages may not sum to 100 because of rounding.
SOURCE: U.S. Department of Education, National Center for Education Statistics, Third International Mathematics and Science Study (TIMSS), Video Study, 1999.

When comparing just two-dimensional geometry problems, the procedural complexity of problems in Japanese lessons looks less different than those in other countries, although there still are a smaller percentage of low complexity problems in Japan than in Australia, Hong Kong SAR, and the Netherlands, and a larger percentage of high complexity problems in Japan than in Australia and Hong Kong SAR.

Mathematical Reasoning

One of the features that distinguishes mathematics from other school subjects is the special forms of reasoning that can be involved in solving problems (National Research Council 2001a). One kind of problem that requires special reasoning is a mathematical proof. To prove that something is true in mathematics means more than inferring it is true by checking a few cases. Rather, it requires demonstrating, through logical argument, that it must be true for all cases. The kind of reasoning required to complete mathematical proofs often is referred to as deductive reasoning. Although the TIMSS 1995 Video Study found that such reasoning did not occur frequently in all the countries (Stigler et al. 1999), it has been recommended as an important aspect of elementary and middle school mathematics (National Council of Teachers of Mathematics 2000; National Research Council 2001a).

All independent and concurrent mathematics problems (see chapter 3) were examined for whether they involved proofs. A problem was coded as a proof if the teacher or students verified or demonstrated that the result must be true by reasoning from the given conditions to the result using a logically connected sequence of steps. For example, in one lesson, students were asked to find the sum of the interior angles of a pentagon [Video clip example 4.4]. The teacher then reviewed the approach recommended in the textbook by demonstrating that a pentagon can be divided into three triangles, that the sum of the interior angles of each triangle is 180 degrees, that the sum of the angles of the three triangles is 540 degrees, and that, therefore, the sum of the angles of a pentagon is 540 degrees. The definition of proof included these rather informal demonstrations because the aim was to capture all problems that included some form of deductive reasoning.

Figures 4.3 and 4.4 show that such problems were evident to a substantial degree only in Japan. On average, 26 percent of the mathematics problems per lesson in Japan included proofs and 39 percent of Japanese eighth-grade mathematics lessons contained at least one proof. These were higher percentages than in the Czech Republic, Hong Kong SAR, and Switzerland. Too few cases of proofs were found in Australia, the Netherlands, and the United States to calculate reliable estimates.

Teaching Mathematics in Seven Countries
Results From the TIMSS 1999 Video Study

FIGURE 4.3. Average percentage of problems per eighth-grade mathematics lesson that included proofs, by country: 1999

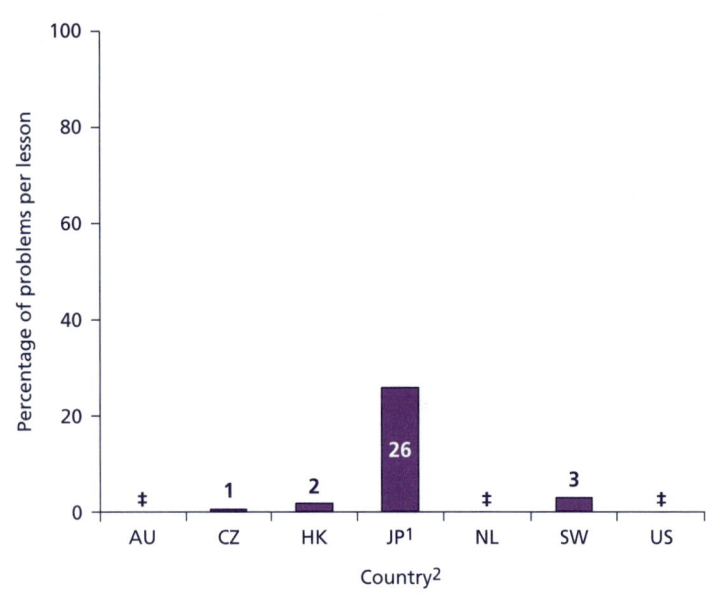

‡Reporting standards not met. Too few cases to be reported.
[1]Japanese mathematics data were collected in 1995.
[2]AU=Australia; CZ=Czech Republic; HK=Hong Kong SAR; JP=Japan; NL= Netherlands; SW=Switzerland; and US=United States.
NOTE: JP>CZ, HK, SW. For each country, average percentage was calculated as the sum of the percentage within each lesson, divided by the number of lessons.
SOURCE: U.S. Department of Education, National Center for Education Statistics, Third International Mathematics and Science Study (TIMSS), Video Study, 1999.

FIGURE 4.4. Percentage of eighth-grade mathematics lessons that contained at least one proof, by country: 1999

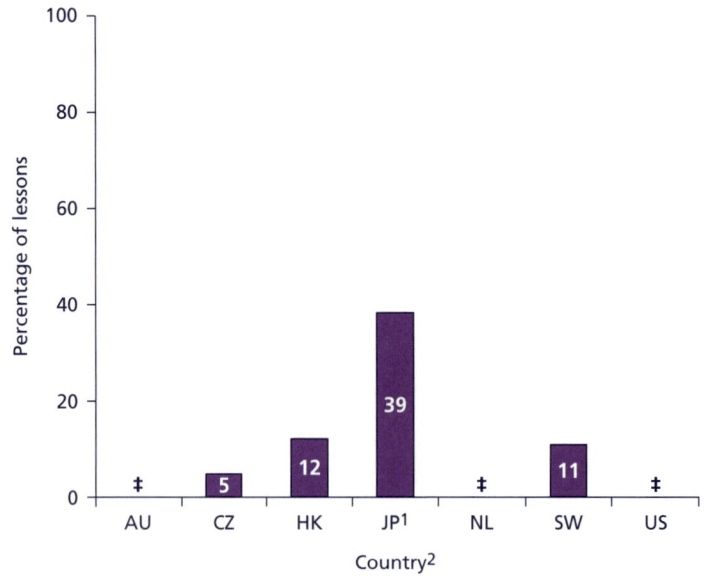

‡Reporting standards not met. Too few cases to be reported.
[1]Japanese mathematics data were collected in 1995.
[2]AU=Australia; CZ=Czech Republic; HK=Hong Kong SAR; JP=Japan; NL= Netherlands; SW=Switzerland; and US=United States.
NOTE: JP>CZ, HK, SW. The percentage reported for Japan differs from that reported in Stigler et al. (1999) because the definition for proof was changed for the current study.
SOURCE: U.S. Department of Education, National Center for Education Statistics, Third International Mathematics and Science Study (TIMSS), Video Study, 1999.

Because many students encounter proofs for the first time when studying two-dimensional geometry, especially when working with polygons, angles, and lines, it is possible that the relatively high frequency of problems including proofs in Japanese eighth-grade lessons was due to the high incidence of these topics in the Japanese sample. Figure 4.5 shows the percentage of two-dimensional geometry problems per lesson in each country coded as proofs. These percentages are based only on lessons that contained two-dimensional geometry problems; therefore, the sample size was much smaller than the full sample. Because of the reduced sample sizes, the results should be interpreted with caution.

FIGURE 4.5. Average percentage of two-dimensional geometry problems that included proofs per eighth-grade mathematics lesson in sub-sample of lessons containing two-dimensional geometry problems, by country: 1999

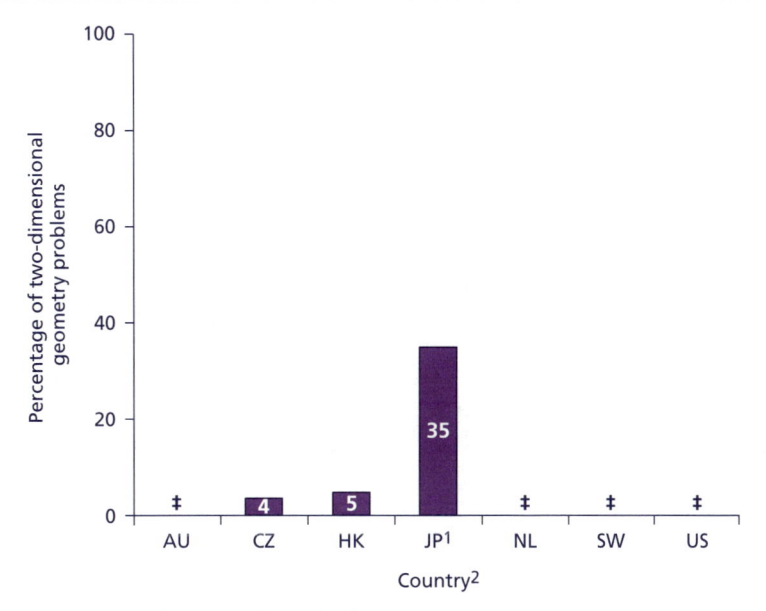

Country	Percentage
AU	‡
CZ	4
HK	5
JP[1]	35
NL	‡
SW	‡
US	‡

‡Reporting standards not met. Too few cases to be reported.
[1]Japanese mathematics data were collected in 1995.
[2]AU=Australia; CZ=Czech Republic; HK=Hong Kong SAR; JP=Japan; NL= Netherlands; SW=Switzerland; and US=United States.
NOTE: JP>CZ, HK. Because the focus of this analysis was on two-dimensional geometry problems, only those lessons that contained such problems were included. Thus, the sample sizes for each country are less than the full sample. Results should therefore be interpreted with caution. For each country, average percentage was calculated as the sum of the percentage within each lesson, divided by the number of lessons.
SOURCE: U.S. Department of Education, National Center for Education Statistics, Third International Mathematics and Science Study (TIMSS), Video Study, 1999.

Controlling for topic had no effect on the relatively high percentage of problems that included proofs in Japanese eighth-grade mathematics lessons compared with those in the other countries where estimates could be reliably reported. In the sample of lessons that contained two-dimensional geometry problems, and in this analysis that considered only two-dimensional geometry problems, a higher percentage of problems in Japanese lessons included proofs than those in the other countries with reliable estimates.

How Mathematics Is Related Over the Lesson

Many factors can influence the clarity and coherence of mathematics lessons. Chapter 3 considered pedagogical factors that can influence the ease with which students identify the main points of the lesson (goal and summary statements) as well as factors that affect the flow of a lesson (lesson interruptions).

The mathematics content itself can contribute to the clarity and coherence of lessons. Because much of the content was carried through the mathematics problems of the lesson, the clarity and coherence of lessons might have been influenced by the way in which the problems within lessons were related to each other.

The mathematics problem analysis group coded the mathematical relationships among all the problems (both independent and concurrent problems, see chapter 3) presented during the lessons. Each problem, except the first problem in the lesson, was classified as one (and only one) of four basic kinds of relationships:[3]

- *Repetition:* The problem was the same, or mostly the same, as a preceding problem in the lesson [Video clip example 4.5]. It required essentially the same operations to solve although the numerical or algebraic expression might be different.

- *Mathematically related:* The problem was related to a preceding problem in the lesson in a mathematically significant way [Video clip example 4.6]. This included using the solution to a previous problem for solving this problem, extending a previous problem by requiring additional operations, highlighting some operations of a previous problem by considering a simpler example, or elaborating a previous problem by solving a similar problem in a different way.

- *Thematically related:* The problem was related to a preceding problem only by virtue of it being a problem of a similar topic or a problem treated under a larger cover story or real-life scenario introduced by the teacher or the curriculum materials. If the problem was mathematically related as well, it was coded only as mathematically related.

- *Unrelated:* The problem was none of the above [Video clip example 4.7]. That is, the problem required a completely different set of operations to solve than previous problems and was not related thematically to any of the previous problems in the lesson.

Mathematically related problems, by definition, tie the content of the lesson together through a variety of mathematical relationships. Sequences of such problems might provide good opportunities for students to construct mathematical relationships and to see the mathematical structure in the topic they are studying (Hiebert et al. 1997; National Research Council 2001a). Repetition problems require little change in students' thinking if students can solve the first problem in the series. These problems often are used for students to practice procedures for solving problems of particular kinds. Unrelated problems, by definition, divide the lesson into mathematically unrelated segments. Figure 4.6 shows the average percentage per eighth-grade mathematics lesson of problems of each kind of relationship.

---

[3]The first problem in each lesson was not coded for a relationship because the coding scheme defined relationship in terms of problems that preceded a given problem, and no problem preceded the first problem.

Chapter 4 | **77**
The Mathematical Content of Lessons

FIGURE 4.6. Average percentage of eighth-grade mathematics problems (excluding the first problem) per lesson related to previous problems, by country: 1999

Country	Unrelated	Repetition	Thematically related	Mathematically related
AU	4	76	8	13
CZ	7	67	11	16
HK	1	69	6	24
JP[1]	‡	40 / 18 (65)	42	22
NL	2	65	12	22
SW	3	73	5	20
US	8	68	9	16

‡Reporting standards not met. Too few cases to be reported.
[1]Japanese mathematics data were collected in 1995.
[2]AU=Australia; CZ=Czech Republic; HK=Hong Kong SAR; JP=Japan; NL= Netherlands; SW=Switzerland; and US=United States.
[3]Mathematically related: HK>AU; JP>AU, CZ, HK, NL, SW, US.
[4]Thematically related: CZ, JP>SW; NL>HK, SW.
[5]Repetition: AU, CZ, HK, NL, SW, US>JP.
[6]Unrelated: CZ>HK, NL, SW.

NOTE: Percentages may not sum to 100 because of rounding. For each country, average percentage was calculated as the sum of the percentage within each lesson, divided by the number of lessons. The tests for significance take into account the standard error for the reported differences. Thus, a difference between averages of two countries may be significant while the same difference between two other countries may not be significant.

SOURCE: U.S. Department of Education, National Center for Education Statistics, Third International Mathematics and Science Study (TIMSS), Video Study, 1999.

Eighth-grade mathematics Japanese lessons contained a higher percentage of problems per lesson (42 percent) that were mathematically related than lessons in all the other countries, and a lower percentage of problems per lesson (40 percent) that were repetitions than all the other countries. Across all the countries, except Japan, at least 65 percent of the problems were repetitions.

Although the percentage of unrelated problems per lesson was relatively small in all countries, it is worth considering these cases further because, by definition, unrelated problems divide the lesson into mathematically unrelated segments. Such breaks in content can interrupt the mathematical flow of a lesson. As seen in figure 4.6, the eighth-grade lessons in the Czech Republic contained a larger percentage of such problems on average than lessons in Hong Kong SAR, the Netherlands, and Switzerland.

The number of unrelated problems provides additional information because the average number of unrelated problems per lesson reveals the average number of content unrelated segments that were contained in a lesson. Lessons in the Czech Republic and the United States contained, on average, more unrelated problems per lesson than lessons in Australia, Hong Kong SAR, the Netherlands, and Switzerland (table 4.2). Based on the fact that the number of mathematically unrelated segments is one more than the number of unrelated problems, table 4.2 shows that in the Czech Republic and the United States, there were on average about two content unrelated segments per eighth-grade mathematics lesson.

TABLE 4.2.	Average number of unrelated problems per eighth-grade mathematics lesson, by country: 1999
Country	Average number of unrelated problems[2]
Australia (AU)	#
Czech Republic (CZ)	1
Hong Kong SAR (HK)	#
Japan[1] (JP)	‡
Netherlands (NL)	#
Switzerland (SW)	#
United States (US)	1

#Rounds to zero.
‡Reporting standards not met. Too few cases to be reported.
[1]Japanese mathematics data were collected in 1995.
[2]CZ, US>AU, HK, NL, SW.
NOTE: For each country, average percentage was calculated as the sum of the percentage within each lesson, divided by the number of lessons.
SOURCE: U.S. Department of Education, National Center for Education Statistics, Third International Mathematics and Science Study (TIMSS), Video Study, 1999.

A question that arises when considering the results in figure 4.6 is whether the differences between the eighth-grade mathematics lessons in Japan and the other countries can be attributed to the topic sample in Japan. As stated earlier, the Japanese sample appeared to contain a high percentage of two-dimensional geometry problems relative to the other countries. To check this, relationships among problems were re-calculated for problems in each country that were in the sub-category of two-dimensional geometry. Because of the relatively small sample sizes, the results should be interpreted with caution. Figure 4.7 shows the results.

Chapter 4
The Mathematical Content of Lessons

FIGURE 4.7. Average percentage of two-dimensional geometry problems (excluding the first problem) related to previous problems per eighth-grade mathematics lesson in sub-sample of lessons containing two-dimensional geometry problems, by country: 1999

[Bar chart showing percentages by country (AU, CZ, HK, JP[1], NL, SW, US) with categories: Mathematically related[3], Thematically related[4], Repetition[5], Unrelated[6]]

Country	Unrelated	Repetition	Thematically related	Mathematically related
AU	14	53	20	13
CZ	20	46	18	16
HK	7	56	12	25
JP[1]	‡	26	32	43
NL	2	56	29	13
SW	8	56	8	28
US	43	31	19	7

‡Reporting standards not met. Too few cases to be reported.
[1]Japanese mathematics data were collected in 1995.
[2]AU=Australia; CZ=Czech Republic; HK=Hong Kong SAR; JP=Japan; NL= Netherlands; SW=Switzerland; and US=United States.
[3]Mathematically related: JP>AU, CZ, NL, US; SW>US.
[4]Thematically related: No difference detected.
[5]Repetition: SW>JP.
[6]Unrelated: CZ>NL.

NOTE: Because the focus of this analysis was on two-dimensional geometry problems, only those lessons that contained such problems were included. Thus, the sample sizes for each country are less than the full sample. Results should therefore be interpreted with caution. Percentages may not sum to 100 because of rounding. For each country, average percentage was calculated as the sum of the percentage within each lesson, divided by the number of lessons. The tests for significance take into account the standard error for the reported differences. Thus, a difference between averages of two countries may be significant while the same difference between two other countries may not be significant.

SOURCE: U.S. Department of Education, National Center for Education Statistics, Third International Mathematics and Science Study (TIMSS), Video Study, 1999.

Controlling for topic with the sample of lessons available shows that Japan retained its relatively high percentage of mathematically related problems (see also figure 4.6). There was no measurable difference detected in the percentage of problems that were repetitions when only two-dimensional geometry problems are considered.

The previous analyses of mathematical relationships within lessons focused on the way in which problems were (or were not) related. Another lens through which to view mathematical relatedness within lessons is to consider mathematical topics and shifts among topics. Parts of lessons can become unrelated when teachers switch from one topic to another. For example, a teacher might review one topic and then introduce new material on another topic. Of course, a teacher might also connect different topics mathematically or thematically. Consequently, shifts between topics do not necessarily produce fragmentation or disjointedness.

To check on topic shifts within lessons, figure 4.8 shows the percentage of eighth-grade mathematics lessons that contained problems related to a single topic. The topics included in this analysis are those shown in table 4.1: the three sub-categories under number, geometry, and algebra, respectively, along with the major topics of statistics and trigonometry.

| FIGURE 4.8. | Percentage of eighth-grade mathematics lessons that contained problems related to a single topic, by country: 1999 |

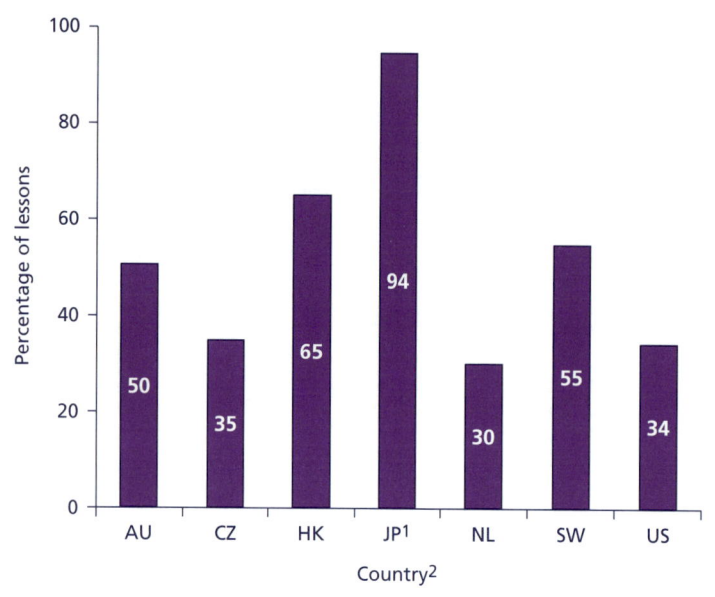

Country	Percentage of lessons
AU	50
CZ	35
HK	65
JP[1]	94
NL	30
SW	55
US	34

[1] Japanese mathematics data were collected in 1995.
[2] AU=Australia; CZ=Czech Republic; HK=Hong Kong SAR; JP=Japan; NL=Netherlands; SW=Switzerland; and US=United States.
NOTE: HK>CZ, NL, US; JP>AU, CZ, HK, NL, SW, US; SW>US. The tests for significance take into account the standard error for the reported differences. Thus, a difference between averages of two countries may be significant while the same difference between two other countries may not be significant.
SOURCE: U.S. Department of Education, National Center for Education Statistics, Third International Mathematics and Science Study (TIMSS), Video Study, 1999.

As shown in figure 4.8, 55 percent or less of the eighth-grade mathematics lessons in five countries contained problems related to a single topic. This means that at least 45 percent of the lessons in these countries included problems related to two or more topics. By comparing this result with the number of unrelated problems per lesson (table 4.2), it is possible to infer that some lessons in most countries contained topic shifts through a sequence of related, rather than unrelated, problems. For example, a shift from operations with linear expressions to solving linear equations can be achieved by solving a related series of problems. This suggests that topic shifts did not necessarily lead to content fragmentation.

Overall, the results on mathematical relationships indicate that, in all the countries, most of the mathematics discussed and studied within these eighth-grade lessons was related. Mathematics lessons, in general, had few if any mathematically disjointed and unrelated segments. For many lessons in most of the countries, however, the relatedness seems to have been achieved, in part, through repetition. Only in Japan were the majority of problems per lesson related mathematically in ways other than repetition (figure 4.6).

Summary

Chapter 3 concluded with an observation that the organizing and structuring features of the eighth-grade mathematics lessons might have enabled some learning opportunities and constrained others. According to this view, the nature of these learning opportunities would be clarified further by considering the nature of the mathematics content presented during the lesson and the way in which the content was treated. This chapter focused on the nature of the content. Some findings reinforce and elaborate the emergent trends in chapter 3, and some findings suggest new images.

- The results in chapter 3 showed that, on average, Japanese eighth-grade mathematics lessons were characterized by devoting lesson time to solving relatively few problems and spending a relatively long time on each one. It appears that this structure was filled with problems possessing a unique content character based on a number of features. Compared with those of all the other countries where reliable estimates could be calculated, the problems in Japanese eighth-grade mathematics lessons were of higher procedural complexity (figure 4.1), they included proofs more often (figures 4.3 and 4.4), and they were related to each other more often in mathematically significant ways (figure 4.6). Follow-up analyses suggest that this profile (except for procedural complexity) cannot be fully explained by the large percentage of two-dimensional geometry problems in the Japanese sample (figures 4.2, 4.5, and 4.7).

- The fact that few differences were found among these countries on several of the variables raises the question of whether eighth-grade teachers in the countries other than Japan teach mathematics in similar ways. No differences were found among the other six countries on the percentage of mathematics problems per lesson that were of high procedural complexity or low procedural complexity (figure 4.1), and no differences were found on the percentage of mathematics problems per lesson that were repetitions (figure 4.6). In addition, no differences were found on the percentage of mathematics problems per lesson that involved proofs among the Czech Republic, Hong Kong SAR, and Switzerland (1, 2, and 3 percent, respectively). Too few mathematics problems that involved proofs were found in Australia, the Netherlands, and the United States to calculate reliable estimates. Do these results point to similar methods of eighth-grade mathematics teaching among the countries other than Japan? This question is explored more fully in chapters 5 and 6.

- The results in this chapter suggest that the purpose segments found in chapter 3 were filled with mathematics problems that were, in general, consistent with the relative emphasis on particular purposes found in several countries. Japan's relative emphasis on introducing new content (chapter 3, figure 3.8) is consistent with the relatively high percentage of mathematically related problems per lesson and relatively low percentage of repetition problems (figure 4.6). Hong Kong SAR's relative emphasis on practicing new content (chapter 3, figure 3.8), and the Czech Republic's and the United States' relative emphasis on review (chapter 3, figure 3.8), are consistent with the relatively large percentage of repetition problems (at least 67 percent) per lesson in these countries (figure 4.6). A large percentage of repetition problems were also found, however, in the other countries—Australia, the Netherlands, and Switzerland. It is reasonable to conjecture that repetition becomes the most common problem-related activity for teaching eighth-grade mathematics unless there is a clear emphasis on introducing new concepts or procedures.

The findings presented in this chapter reveal some of the similarities and differences among countries in the content of eighth-grade mathematics lessons. Additional descriptions of lesson content, generated by the mathematics quality analysis group (see appendix A), are presented in the experimental analyses in appendix D.

The importance of the mathematics content presented during the lesson derives, in part, from the fact that content defines the parameters within which students work. If students are presented with a topic, they have an opportunity to learn something about the topic. If the topic is not introduced, there is little chance students will learn about it, at least in school. However, the fact that a topic has been introduced does not say much about quality of the learning opportunity or about how deeply students might learn the topic. To pursue these issues, it is important to know how the content was worked on during the lesson. In what context were the problems presented? Did they invite exploration by the students or were they simply exercises in executing procedures? What kind of mathematics work did students do when they worked on their own? Answers to these kinds of questions, taken up in the next chapter, provide additional information about eighth-grade mathematics teaching in each country.

CHAPTER 5
Instructional Practices: How Mathematics Was Worked On

The way in which mathematics content was worked on during the lesson adds important information about the learning opportunities for students. In what kind of context were mathematical problems embedded? Did students have a choice in the methods they used to solve the problems? What were students expected to do when they were working on their own? Answers to these kinds of questions add key elements to the pictures emerging from chapters 3 and 4.

Recall that chapter 3 focused on the way in which lessons were structured and chapter 4 described the nature of the mathematical content. These aspects of lessons help shape the learning opportunities for students. An eighth-grade mathematics lesson that introduces new concepts or procedures, for example, would seem to provide different learning opportunities than a lesson that reviews material students already have studied (chapter 3). Similarly, a lesson that engages students in constructing mathematical proofs, for example, would seem to provide different learning opportunities than a lesson that asks students to practice executing procedures on a set of similar problems (chapter 4). In brief, lesson structure provides a framework within which learning opportunities are created, and the mathematical content helps to define and set additional boundaries on these opportunities.

This chapter fills in additional information by exploring four different aspects of instructional practices that were used during the lessons. The first, and most central of the four, is the way in which mathematical problems were presented and solved. Because so much of the mathematics instruction in all the countries was carried through presenting and solving problems (see chapter 3, figure 3.5), it is useful to examine this activity in more detail. The chapter also describes what happened during the non-problem segments of lessons, presents summary indicators of the discourse in the lessons, and identifies the instructional resources used during the lessons.

How Mathematical Problems Were Presented and Worked On

How were mathematical problems presented and how were they solved? The international coding team, the mathematics problem analysis group, and the problem implementation analysis group (see appendix A for descriptions) explored various aspects of presenting and solving problems, including the following:

- *The context in which problems were presented and solved:* Whether problems were connected with real-life situations, whether representations were used to present the information, whether physical materials were used, and whether the problems were applications (i.e., embedded in verbal or graphic situations).

- *Specific features of how problems were worked on during the lesson:* Whether a solution to the problem was stated publicly, whether alternative solution methods were presented, whether students had a choice in the solution method they used, and whether teachers summarized the important points after problems were solved.

- *The kind of mathematical processes that were used to solve problems:* What kinds of processes were made visible for students during the lesson and what kinds were used by students when working on their own.

The majority of investigations discussed in this chapter applied to all mathematics problems in all the lessons, both publicly discussed and privately worked on, excluding answered-only problems.[1] Many codes were not applied to answered-only problems because, by definition, they were not worked on in the filmed lesson. Most of the analyses described below were conducted on the complete set of independent and concurrent problems combined (see chapter 3 for definitions). Where analyses on subgroups of these problems or countries were conducted, a rationale is provided.

Problem Context

Real-life situations

Mathematical problems can be presented to students within a real-life context [Video clip example 5.1] or by using only mathematical language with written symbols [Video clip example 5.2]. "Estimate the surface area of the frame in the picture below," and "Samantha is collecting data on the time it takes her to walk to school. A table shows her travel times over a two-week period; find the mean," are examples of real-life contexts. "Graph the equation: $y = 3x + 7$," and "Find the volume of a cube whose side measures 3.5 cm," are examples of problems presented only with mathematical language.

The appropriate relationship of mathematics to real life has been discussed for a long time (Davis and Hersh 1981; Stanic and Kilpatrick 1988). Some psychologists and mathematics educators have argued that emphasizing the connections between mathematics and real-life situations can distract students from the important ideas and relationships within mathematics (Brownell 1935; Prawat 1991). Others have claimed some significant benefits of presenting mathematical problems in the context of real-life situations, including that such problems connect better with students' intuitions about mathematics, they are useful for showing the relevance of mathematics, and they are more interesting for students (Burkhardt 1981; Lesh and Lamon 1992; Streefland 1991).

Figure 5.1 shows the percentage of problems per eighth-grade mathematics lesson that were presented or set up using real-life situations. If teachers brought in real-life connections later, when solving the problems, this was marked separately.

---

[1] Answered-only problems had already been completed prior to the videotaped lesson, and only their answers were shared. They included no public discussion of a solution procedure and no time in which students worked on them privately (see chapter 3).

Chapter 5
Instructional Practices: How Mathematics Was Worked On

FIGURE 5.1. Average percentage of problems per eighth-grade mathematics lesson that were either set up with the use of a real-life connection, or set up using mathematical language or symbols only, by country: 1999

Country	Set-up contained a real-life connection[3]	Set-up used mathematical language or symbols only[4]
AU	27	72
CZ	15	81
HK	15	83
JP[1]	9	89
NL	42	40
SW	25	71
US	22	69

[1] Japanese mathematics data were collected in 1995.
[2] AU=Australia; CZ=Czech Republic; HK=Hong Kong SAR; JP=Japan; NL=Netherlands; SW=Switzerland; and US=United States.
[3] Set-up contained a real-life connection: AU, SW>JP; NL>CZ, HK, JP, US.
[4] Set-up used mathematical language or symbols only: AU, CZ, HK, JP, SW, US>NL; JP>AU, SW, US.

NOTE: Analyses do not include answered-only problems (i.e., problems that were completed prior to the videotaped lesson and only their answers were shared). Percentages may not sum to 100 because some problems were marked as "unknown" and are not included here. For each country, average percentage was calculated as the sum of the percentage within each lesson, divided by the number of lessons.
SOURCE: U.S. Department of Education, National Center for Education Statistics, Third International Mathematics and Science Study (TIMSS), Video Study, 1999.

In the Netherlands, a smaller percentage of problems were set up using mathematical language or symbols only (40 percent of problems per lesson, on average) than in any other country. In the other six countries, between 69 percent and 89 percent of the problems were set up only with numbers and symbols. In Japan, 89 percent of problems were set up by using mathematical language or symbols only. That figure was greater than the percentages in Australia, the Netherlands, Switzerland, and the United States.

The other choice for setting up problems that teachers made—using real-life connections—shows, of course, a nearly reverse set of country differences. A higher percentage of problems per lesson, on average, in the Netherlands were set up with a real-life connection (42 percent) than in the Czech Republic, Hong Kong SAR, Japan, and the United States (ranging from 9 percent to 22 percent).

In all the countries, if teachers made real-life connections, they did so at the initial presentation of the problem rather than only while solving the problem. A small percentage of eighth-grade mathematics lessons were taught by teachers who introduced a real-life connection to help solve the problem if such a connection had not been made while presenting the problem (from less than 1 to 3 percent for all the countries).

Representations

Another contextual variable is the representation of the mathematical information of a problem. Representations usually include numerals and other conventional written symbols, but they also can include drawings or diagrams, tables, and graphs. Figure 5.2 shows the percentage of problems per eighth-grade mathematics lesson that contained these forms of representation. To count as a drawing or diagram, a figure must have included information relevant for solving the problem. Excluded were motivational diagrams that lacked such information (e.g., a photo of an Olympic runner that accompanied a story problem on race times). A table was defined as an arrangement of numbers, signs, or words that exhibited a set of facts or relations in a definite, compact, and comprehensive form. Typically, a table contained rows and/or columns that were labeled and had borders. Graphs included statistical displays such as bar graphs and line graphs.

FIGURE 5.2. Average percentage of problems per eighth-grade mathematics lesson that contained a drawing/diagram, table, and/or graph, by country: 1999

Country	Drawing/diagram[3]	Table[4]	Graph[5]
AU	29	18	5
CZ	22	6	4
HK	30	12	6
JP[1]	83	4	8
NL	25	15	13
SW	35	18	5
US	26	12	14

[1] Japanese mathematics data were collected in 1995.
[2] AU=Australia; CZ=Czech Republic; HK=Hong Kong SAR; JP=Japan; NL=Netherlands; SW=Switzerland; and US=United States.
[3] Drawing/diagram: JP>AU, CZ, HK, NL, SW, US.
[4] Table: NL, SW>CZ, JP.
[5] Graph: No differences detected.

NOTE: Analyses do not include answered-only problems (i.e., problems that were completed prior to the videotaped lesson and only their answers were shared). For each country, average percentage was calculated as the sum of the percentage within each lesson, divided by the number of lessons. The tests for significance take into account the standard error for the reported differences. Thus, a difference between averages of two countries may be significant while the same difference between two other countries may not be significant. Differences found among countries in the use of drawings/diagrams may be because of the focus on two-dimensional geometry in the Japanese sample of lessons.
SOURCE: U.S. Department of Education, National Center for Education Statistics, Third International Mathematics and Science Study (TIMSS), Video Study, 1999.

The most noticeable difference among countries is that Japanese lessons contained, on average, a larger percentage of problems with drawings or diagrams than did eighth-grade mathematics lessons in the other countries. As seen in figure 5.2, drawings/diagrams were present in roughly 20 to 35 percent of problems across all the countries except Japan, where they were included in about 80 percent of problems.

Drawings and diagrams are often used by teachers when working with problems of two-dimensional geometry. The results therefore suggest that the difference between Japan and other countries was associated with the apparent higher frequency of two-dimensional geometry problems in the Japanese data set (see chapter 4, figure 4.1). Analyzing the sub-sample of lessons that contained two-dimensional geometry problems showed no detectable differences on the percentage of problems per lesson that used drawings/diagrams (data not shown in table or figure). When taking into account only two-dimensional geometry problems, differences between Japan and the other countries that were found in figure 5.2 were not detectable.

The percentage of two-dimensional geometry problems per eighth-grade mathematics lesson, on average, that contained a drawing or diagram ranged from 60 percent (in the United States) to 94 percent (in Japan and Switzerland). These percentages are based only on lessons that contained two-dimensional geometry problems; therefore, the sample size was much smaller than the full sample. Because of the reduced sample sizes, the results should be interpreted with caution.

Though used in no more than 18 percent of problems in any of the countries, tabular representations were more frequently part of the set-up or solution process of problems in the Netherlands and Switzerland than they were in the Czech Republic and Japan. The percentage of mathematical problems that included graphical representations in the set-up or solution process was not found to differ across the countries.

Physical materials

Incorporating physical materials in the teaching of mathematics has a long tradition in mathematics education. Specially designed materials can be used to illustrate mathematical objects or relationships or they can serve as instruments to measure quantities. Tangrams, for example, can be used to explore different ways in which surface areas can be covered. The research evidence suggests that physical materials can, but do not necessarily, facilitate improved student learning (National Research Council 2001a).

Figure 5.3 shows the percentage of problems per eighth-grade mathematics lesson that involved the use of physical materials. Physical materials included measuring instruments (e.g., rulers, protractors, compasses), special mathematical materials (e.g., tiles, tangrams, base-ten blocks), geometric solids, and cut-out plane figures. Papers, pencils, calculators, and computers were not included in this analysis. To be counted, the materials must have been used or manipulated by the teacher or student(s) when presenting or solving the problem, not simply present in the classroom [Video clip example 5.3].

Teaching Mathematics in Seven Countries
Results From the TIMSS 1999 Video Study

FIGURE 5.3.	Average percentage of problems per eighth-grade mathematics lesson that involved the use of physical materials, by country: 1999

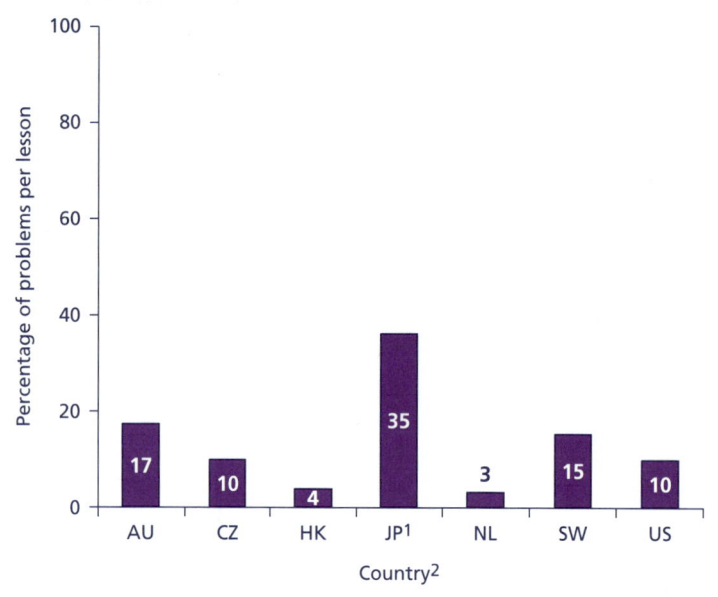

Bar chart showing percentage of problems per lesson by country:
- AU: 17
- CZ: 10
- HK: 4
- JP[1]: 35
- NL: 3
- SW: 15
- US: 10

[1] Japanese mathematics data were collected in 1995.
[2] AU=Australia; CZ=Czech Republic; HK=Hong Kong SAR; JP=Japan; NL=Netherlands; SW=Switzerland; and US=United States.
NOTE: AU, SW>HK, NL; JP>CZ, HK, NL, US. Analyses do not include answered-only problems (i.e., problems that were completed prior to the videotaped lesson and only their answers were shared). For each country, average percentage was calculated as the sum of the percentage within each lesson, divided by the number of lessons.
SOURCE: U.S. Department of Education, National Center for Education Statistics, Third International Mathematics and Science Study (TIMSS), Video Study, 1999.

A larger percentage of problems per eighth-grade mathematics lesson in Japan (35 percent) included the use of physical materials than the percentages in the Czech Republic, Hong Kong SAR, the Netherlands, and the United States (10, 4, 3, and 10 percent, respectively).

Because geometry problems lend themselves to using specially designed materials, it is possible that the relatively large average percentage of problems per lesson in Japan that included physical materials might be associated with the apparently high percentage of problems per lesson in the Japanese sample that focused on two-dimensional geometry (see chapter 4, figure 4.1). Figure 5.4 shows the results of analyzing the sub-sample of lessons in each country that contained two-dimensional geometry problems for use of physical materials. The sample for this analysis was much smaller than the full sample of eighth-grade mathematics lessons. Because of the relatively small sample sizes, the results should be interpreted with caution.

FIGURE 5.4. Average percentage of two-dimensional geometry problems per eighth-grade mathematics lesson that involved the use of physical materials, by country: 1999

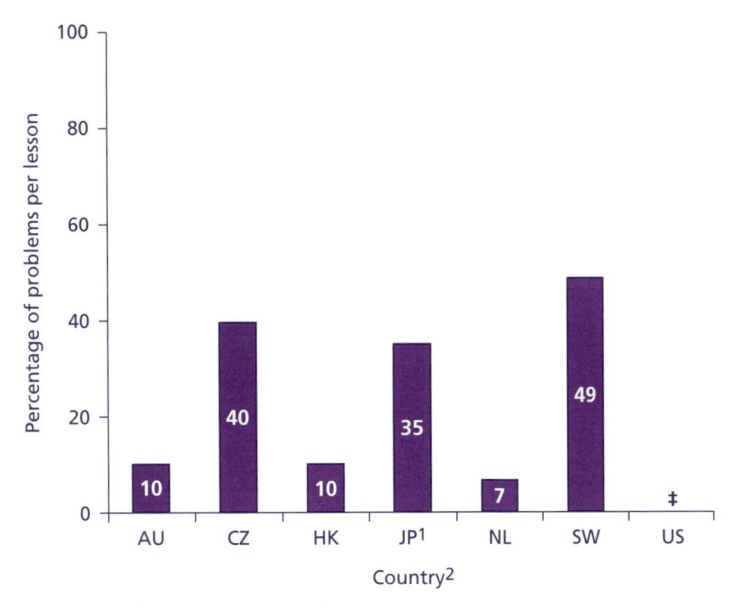

[Bar chart showing percentage of problems per lesson by country:
AU: 10, CZ: 40, HK: 10, JP[1]: 35, NL: 7, SW: 49, US: ‡]

‡Reporting standards not met. Too few cases to be reported.
[1]Japanese mathematics data were collected in 1995.
[2]AU=Australia; CZ=Czech Republic; HK=Hong Kong SAR; JP=Japan; NL=Netherlands; SW=Switzerland; and US=United States.
NOTE: CZ>HK, NL; JP, SW>AU, HK, NL. Because the focus of this analysis was on two-dimensional geometry problems, only those lessons that contained such problems were included. Thus, the sample sizes for each country are less than the full sample. Results should therefore be interpreted with caution. Analyses do not include answered-only problems (i.e., problems that were completed prior to the videotaped lesson and only their answers were shared). For each country, average percentage was calculated as the sum of the percentage within each lesson, divided by the number of lessons.
SOURCE: U.S. Department of Education, National Center for Education Statistics, Third International Mathematics and Science Study (TIMSS), Video Study, 1999.

When considering physical material use only in two-dimensional geometry problems compared to the full sample of problems, the percentages in several countries appear to be different. The pairwise comparisons show several changes in relative differences among countries. In Japanese and Swiss lessons that included two-dimensional geometry problems, on average 35 percent and 49 percent of the problems per lesson, respectively, involved physical materials. These are larger percentages of problems per lesson compared to Australia, Hong Kong SAR, and the Netherlands. Two-dimensional geometry lessons in the Czech Republic also contained a larger percentage of problems that involved physical materials (40 percent) than such lessons in Hong Kong SAR and the Netherlands.

Because the research evidence suggests that students' learning is influenced by how physical materials are used, not just whether they are used (National Research Council 2001a), it is impossible to draw conclusions about the way in which more frequent use of physical materials in some countries impacted students' learning opportunities. Yet, the fact that eighth-grade mathematics lessons in these countries differed with respect to the frequency with which physical materials were used to solve problems suggests that countries make different choices about incorporating these materials. The apparent changes in percentages and between-country comparisons when considering only two-dimensional geometry problems also suggest that the use of materials was associated with the mathematical topics being taught.

Applications

Working on mathematical problems can take a variety of forms. For example, students can be taught a particular procedure and then asked to practice that procedure on a series of similar problems. These problems can be called exercises. Alternatively, students can be asked to apply procedures they have learned in one context in order to solve problems presented in a different context. These problems can be called applications [Video clip example 5.4]. Applications often are presented using verbal descriptions, graphs, or diagrams rather than just mathematical symbols. They are important because they require students to make decisions about how and when to use procedures they may have already learned and practiced. In this sense, applications are, by definition, more conceptually demanding than routine exercises for the same topic.

Applications might, or might not, be presented in real-life settings. Three examples of application problems are the following.

- "The sum of three consecutive integers is 240. Find the integers." To solve the problem, students might use a guess-and-check method or they might use what they have studied about solving linear equations to represent the situation with the equation $x + (x + 1) + (x + 2) = 240$, and then solve the equation to find the three integers.

- "A rectangular garden is twice as long as it is wide. If the length of the fence enclosing the garden is 24 meters, what are the dimensions of the garden?" Students might use an algebraic representation for this problem as well: $w + 2w + w + 2w = 24$.

- "Find the measure of angle x as shown in figure 5.5." Students might use procedures for finding the sum of angles in a triangle and for finding supplementary angles to find the measure of angle x.

FIGURE 5.5. Example of an application problem: "Find the measure of angle x."

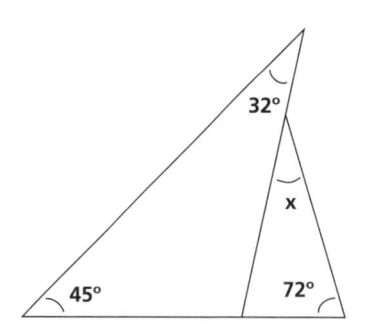

SOURCE: U.S. Department of Education, National Center for Education Statistics, Third International Mathematics and Science Study (TIMSS), Video Study, 1999.

The mathematical problem analysis group classified the problems in the full sample either as applications or exercises. Figure 5.6 shows the percentages of problems per lesson, on average, classified as applications. Japanese lessons contained a higher percentage of applications per lesson (74 percent) than did eighth-grade mathematics lessons from all the other countries except Switzerland (55 percent).

Chapter 5
Instructional Practices: How Mathematics Was Worked On

FIGURE 5.6. Average percentage of problems per eighth-grade mathematics lesson that were applications, by country: 1999

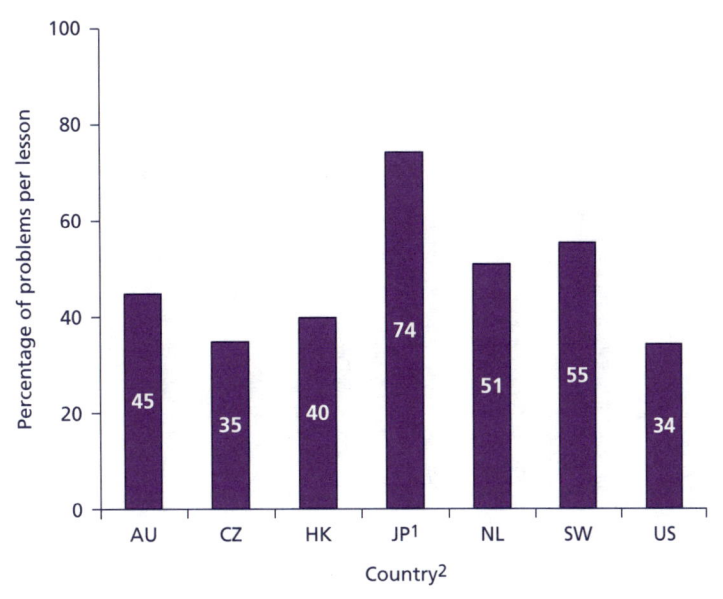

[Bar chart showing Percentage of problems per lesson by Country:
- AU: 45
- CZ: 35
- HK: 40
- JP[1]: 74
- NL: 51
- SW: 55
- US: 34]

[1] Japanese mathematics data were collected in 1995.
[2] AU=Australia; CZ=Czech Republic; HK=Hong Kong SAR; JP=Japan; NL=Netherlands; SW=Switzerland; and US=United States.
NOTE: JP>AU, CZ, HK, NL, US; NL>US; SW>CZ. Analyses do not include answered-only problems (i.e., problems that were completed prior to the videotaped lesson and only their answers were shared). For each country, average percentage was calculated as the sum of the percentage within each lesson, divided by the number of lessons. The tests for significance take into account the standard error for the reported differences. Thus, a difference between averages of two countries may be significant while the same difference between two other countries may not be significant.
SOURCE: U.S. Department of Education, National Center for Education Statistics, Third International Mathematics and Science Study (TIMSS), Video Study, 1999.

Solutions Presented Publicly

One way to measure the extent to which mathematics work was a public versus a private activity during the lessons is to ask whether the answers to mathematics problems were presented publicly. Public presentation of solutions suggests that the whole class was working on the same problem and allows the possibility that the teacher and students might discuss the problem. On the other hand, no public presentation means that students were expected to complete the problem privately, with no follow-up discussion during the lesson. Different students might, or might not, be solving the same problems.

Figure 5.7 shows the percentage of problems per eighth-grade mathematics lesson whose solutions were presented publicly. Independent and concurrent problems were examined separately because concurrent problems were often not discussed publicly at all during the videotaped lesson. Indeed, lessons might have contained a large number of concurrent problems to be completed by students over the span of several days.

FIGURE 5.7. Average percentage of problems per eighth-grade mathematics lesson for which a solution was presented publicly in the videotaped lesson, by country: 1999

■ Independent problems for which a solution was presented publicly[3]

□ Concurrent problems for which a solution was presented publicly[4]

Country	Independent	Concurrent
AU	91	38
CZ	94	76
HK	94	61
JP[1]	88	61
NL	95	16
SW	91	40
US	91	43

[1] Japanese mathematics data were collected in 1995.
[2] AU=Australia; CZ=Czech Republic; HK=Hong Kong SAR; JP=Japan; NL=Netherlands; SW=Switzerland; and US=United States.
[3] Independent problems for which a solution was presented publicly: No differences detected.
[4] Concurrent problems for which a solution was presented publicly: AU, CZ, HK, JP, SW, US>NL; CZ>AU, SW, US; HK>AU, SW.

NOTE: Independent problems were presented as single problems and worked on for a clearly definable period of time. Concurrent problems were presented as a set of problems to be worked on privately. For each country, average percentage was calculated as the sum of the percentage within each lesson, divided by the number of lessons.

SOURCE: U.S. Department of Education, National Center for Education Statistics, Third International Mathematics and Science Study (TIMSS), Video Study, 1999.

Across all the countries, nearly all independent problems included the public presentation of a solution, ranging from 88 percent to 95 percent of independent problems per lesson, on average. No detectable differences were found among the countries. This finding is not surprising because independent problems were defined as problems presented individually, worked on for a clearly definable period of time, with the possibility that they would be solved through a whole class activity (see chapter 3).

The story for concurrent problems appears different, however. On average, 16 percent of concurrent problems per eighth-grade mathematics lesson in the Netherlands included the public presentation of a solution, a lower percentage than in any other country. In contrast, 76 percent of concurrent problems per lesson in the Czech Republic reached a public solution. The Czech average percentage per lesson was significantly greater than the Australian (38 percent), Swiss (40 percent), and United States (43 percent) averages. Concurrent problems, by definition, were presented as a set, to be worked on privately by the students. The fact that Dutch teachers retained the private nature of these problems to a greater extent than teachers in other countries seems to be consistent with the comparatively greater percentage of time devoted to private work in Dutch lessons (figure 3.10). This indicator strengthens the emerging picture of the relatively greater emphasis that Dutch teachers placed on the independent, private work of students compared with public, whole-class work.

Developing and Discussing Solution Methods

Eighth-grade students in all the countries spent at least 80 percent of their time per lesson, on average, solving mathematical problems (chapter 3, figure 3.3). In solving problems, key learning opportunities are created by the way in which methods for solving problems are developed and discussed (Hiebert et al. 1996; Schoenfeld 1985). Thus, to understand the nature of the learning opportunities available to students in these mathematics lessons, it is important to examine the way in which methods for solving problems were treated.

One way of treating the solution methods for mathematics problems is for the teacher to demonstrate one method for solving the problem and then for the students to practice the method on similar problems. This is a common approach in the United States (Fey 1979; Stigler and Hiebert 1999), as well as in other countries (Leung 1995). However, there are some compelling theoretical arguments, along with some empirical data, to suggest that students can benefit from both examining alternative solution methods and being allowed some choice in how they solve the problem (Brophy 1999; National Research Council 2001a). The results presented below were obtained by identifying problems for which more than one solution method was presented and for which students participated in developing the solution methods.

Were alternative solution methods presented publicly?

A class discussion about alternative solution methods for a problem requires, at least, that more than one solution method be presented publicly. To check the frequency of such an event, it was necessary to define alternative solution methods and, in the process, to define what is meant by a solution method. A solution method was defined as a sequence of mathematical steps used to produce a solution. Solution methods could be presented in written or verbal form solely by the teacher, worked out collaboratively with students, or presented solely by students. To count as an alternative solution method, each method needed to (1) be distinctly different from other methods presented, (2) have enough detail so that an attentive student could follow the steps and use the method to produce a solution, and (3) be accepted by the teacher as a distinct and legitimate method, rather than as a correction or elaboration of another method [Video clip example 5.5].

Table 5.1 shows that in all the countries, except Japan, 5 percent or fewer of the problems per eighth-grade mathematics lesson, on average, included the public presentation of alternative solution methods. In Japan, 17 percent of the problems per lesson included alternative methods. The percentage for Japan was higher than that for Australia (2 percent), the Czech Republic (2 percent), and Hong Kong SAR (4 percent). When the countries were compared on the number of lessons that contained at least one example of a problem with more than one solution method, the only detectable difference was that the United States (37 percent of lessons) was greater than the Czech Republic (16 percent of lessons).

TABLE 5.1. Average percentage of problems per eighth-grade mathematics lesson and percentage of lessons with at least one problem in which more than one solution method was presented publicly, by country: 1999

Country	Average percentage of problems per lesson with more than one solution method presented[2]	Percentage of lessons with at least one problem in which more than one solution method was presented[3]
Australia (AU)	2	25
Czech Republic (CZ)	2	16
Hong Kong SAR (HK)	4	23
Japan[1] (JP)	17	42
Netherlands (NL)	5	30
Switzerland (SW)	4	24
United States (US)	5	37

[1] Japanese mathematics data were collected in 1995.
[2] Average percentage of problems per lesson with more than one solution method presented: JP>AU, CZ, HK.
[3] Percentage of lessons with at least one problem in which more than one solution method was presented: US>CZ.
NOTE: Analyses do not include answered-only problems (i.e., problems that were completed prior to the videotaped lesson and only their answers were shared). For each country, average percentage was calculated as the sum of the percentage within each lesson, divided by the number of lessons. The tests for significance take into account the standard error for the reported differences. Thus, a difference between averages of two countries may be significant while the same difference between two other countries may not be significant.
SOURCE: U.S. Department of Education, National Center for Education Statistics, Third International Mathematics and Science Study (TIMSS), Video Study, 1999.

Were students asked to choose their own solution method?

Teachers might expect students to follow a prescribed method when solving a problem, or they might ask or allow students to decide how they would like to solve it. This issue is not entirely the same as presenting multiple solution methods, because even if only one solution method was publicly presented, students might still have had a choice about how to solve the problem.

Student choice was marked when either of the following events occurred: (1) the teacher (or textbook) explicitly stated that students were allowed to use whatever method they wished to solve the problem or (2) two or more solution methods were identified and students were explicitly asked to choose one of the identified methods [Video clip example 5.6]. This analysis provides a conservative estimate of occurrence because there might have been an unspoken understanding in the classroom that students were free to choose their own solution methods. If such cases occurred, they were not included in the problems identified as allowing a choice of solution methods.

Table 5.2 shows that 9 percent or less of problems per eighth-grade mathematics lesson in all the countries except Japan (15 percent) were accompanied with a clear indication that students could select their own solution method. Table 5.2 also shows that the practice of offering students a choice of solution methods, at least on one problem, occurred in a greater percentage of lessons in the United States than in the Czech Republic and Hong Kong SAR. Because student choice on only one problem was sufficient to include the lesson in this analysis, it is an indication of how broadly it occurred, not how frequently it occurred within a lesson.

TABLE 5.2. Average percentage of problems per eighth-grade mathematics lesson and percentage of lessons with at least one problem in which students had a choice of solution methods, by country: 1999

Country	Average percentage of problems per lesson in which students had a choice of solution methods[2]	Percentage of lessons with at least one problem in which students had a choice of solution methods[3]
Australia (AU)	8	25
Czech Republic (CZ)	4	20
Hong Kong SAR (HK)	3	17
Japan[1] (JP)	15	31
Netherlands (NL)	‡	‡
Switzerland (SW)	7	24
United States (US)	9	45

‡Reporting standards not met. Too few cases to be reported.
[1]Japanese mathematics data were collected in 1995.
[2]Average percentage of problems per lesson in which students had a choice of solution methods: no differences detected.
[3]Percentage of lessons with at least one problem in which students had a choice of solution methods: US>CZ, HK.
NOTE: Analyses do not include answered-only problems (i.e., problems that were completed prior to the videotaped lesson and only their answers were shared). For each country, average percentage was calculated as the sum of the percentage within each lesson, divided by the number of lessons.
SOURCE: U.S. Department of Education, National Center for Education Statistics, Third International Mathematics and Science Study (TIMSS), Video Study, 1999.

Did students participate in presenting and examining alternative solution methods?

Tables 5.1 and 5.2 indicate the extent to which alternative solution methods were presented publicly and how often students were explicitly permitted to use a method of their choice. What is not yet clear is whether eighth-grade students were actively involved in developing and examining alternative solution methods—referred to here as "examining methods" [Video clip example 5.7]. Given the percentages in tables 5.1 and 5.2, it appears that, if such an activity did occur, it was a relatively rare event because such instances would have been a subset of the two previous results.

In fact, the results in table 5.3 confirm that such an activity was rare, occurring in 3 percent or less of problems per eighth-grade mathematics lesson in all the countries except Japan, where it occurred with 9 percent of problems per lesson. The rate of occurrence in Japan was higher than in Australia, the Czech Republic, Hong Kong SAR, and the United States. An examination of the lessons with at least one such problem shows that the average percentage in the Czech Republic was less than in Japan and the United States.

It should be noted that the criteria for inclusion as "examining methods" were set quite high: problems were required to include (1) student choice of solution methods, (2) alternative solution methods presented publicly, (3) at least one solution method presented by a student, and (4) a critique or extended examination of a particular method or a comparison of solution methods. Although this kind of classroom activity has been described in the mathematics education literature (Hiebert et al. 1997; Lampert 2001; National Council of Teachers of Mathematics 2000; National Research Council 2001a; Schifter and Fosnot 1993), it was not a frequent part of the eighth-grade mathematics lessons in these countries.

TABLE 5.3. Average percentage of "examining methods" problems per eighth-grade mathematics lesson and percentage of lessons with at least one "examining methods" problem, by country: 1999

Country	Average percentage of examining methods problems[2]	Percentage of lessons with at least one examining methods problem[3]
Australia (AU)	1	8
Czech Republic (CZ)	#	3
Hong Kong SAR (HK)	1	12
Japan[1] (JP)	9	24
Netherlands (NL)	‡	‡
Switzerland (SW)	3	14
United States (US)	2	17

#Rounds to zero.
‡Reporting standards not met. Too few cases to be reported.
[1]Japanese mathematics data were collected in 1995.
[2]Average percentage of examining methods problems: JP>AU, CZ, HK, US.
[3]Percentage of lessons with at least one examining methods problem: JP, US>CZ.
NOTE: Analyses do not include answered-only problems (i.e., problems that were completed prior to the videotaped lesson and only their answers were shared). For each country, average percentage was calculated as the sum of the percentage within each lesson, divided by the number of lessons. "Examining methods" problems were required to include (1) student choice of solution methods, (2) alternative solution methods presented publicly, (3) at least one solution method presented by a student, and (4) a critique or extended examination of a particular method or a comparison of solution methods.
SOURCE: U.S. Department of Education, National Center for Education Statistics, Third International Mathematics and Science Study (TIMSS), Video Study, 1999.

Problem Summaries

After a problem has been solved, teachers might summarize the mathematical points that the problem illustrates. This is one way of clarifying for students what they have just learned by solving the problem or what mathematical concepts or procedures are important to remember for future work. A problem was counted as including a summary if the teacher (or, on rare occasions, a student) restated the major steps used in the solution method or drew attention to a critical mathematical rule or property in the problem [Video clip example 5.8]. The summary must have been provided after the solution was reached. All independent problems were included in this analysis along with concurrent problems for which a solution was stated publicly. Note that problem summaries are different from lesson summaries, presented in chapter 3 (figure 3.13).

Table 5.4 shows that in Japanese eighth-grade mathematics lessons, a higher percentage of problems per lesson were summarized by the teacher (27 percent, on average) compared to lessons from the other countries. Lessons in the Netherlands included a smaller percentage of summarized problems (5 percent, on average) than did those in the Czech Republic, Hong Kong SAR, Japan, and Switzerland.

TABLE 5.4. Average percentage of problems per eighth-grade mathematics lesson that were summarized, by country: 1999

Country	Average percentage of problems per lesson that were summarized
Australia (AU)	9
Czech Republic (CZ)	11
Hong Kong SAR (HK)	13
Japan[1] (JP)	27
Netherlands (NL)	5
Switzerland (SW)	13
United States (US)	6

[1] Japanese mathematics data were collected in 1995.
NOTE: CZ, HK>NL; HK>US; JP>AU, CZ, HK, NL, SW, US; SW>NL, US. Analyses do not include answered-only problems (i.e., problems that were completed prior to the videotaped lesson and only their answers were shared). Analyses do not include concurrent problems (i.e., problems presented as a set to be worked on privately) for which a solution was not publicly presented. For each country, average percentage was calculated as the sum of the percentage within each lesson, divided by the number of lessons.
SOURCE: U.S. Department of Education, National Center for Education Statistics, Third International Mathematics and Science Study (TIMSS), Video Study, 1999.

Mathematical Problems Stated and Solved

A different perspective that can be applied to the presenting and solving of mathematical problems is to compare the nature of the problem statements with the way in which the problems are publicly solved. Previous research has shown that problem statements can be examined for the nature of the mathematical work that is implied and then compared with the mathematical work that actually is performed—and made explicit for the students—while the problems are being solved (Stein, Grover, and Henningsen 1996; Stein and Lane 1996; Smith 2000). The statements of problems imply that particular kinds of mathematical processes will be engaged, but when teachers work through problems, the kinds of processes that students actually engage in or see others use might be different.

Suppose, for example, that the following problem is presented: "Solve for x in the equation $2x + 3 = 11$." The problem statement suggests that a procedure will be used to find x, perhaps subtracting 3 from both sides of the equation and then dividing both sides by 2, yielding $x = 4$. If the problem actually is solved in this way (by the teacher or the students) without further examination, the mathematical processes suggested by the statement and those used while solving the problem are the same. The processes could be called "using procedures."

But imagine that the teacher asks some additional questions as the problem is being solved: "If the equation was written $11 = 2x + 3$, would the solution be the same?" or "Is it OK to divide both sides of the equation by any number?" And imagine that the teacher follows the questions with a discussion on, for example, the concept of transforming equations in ways that preserve equivalence. In this case, the problem statement would have suggested processes of using procedures but the solving activity actually involved analyzing concepts and making connections among mathematical ideas. So, the mathematical processes suggested by the problem statement would not have matched those actually employed when solving it.

The mathematical processes used when solving problems appear to shape the kind of learning opportunities available for students and have been shown to influence the nature of students' learning outcomes (Stein and Lane 1996). Consequently, this perspective provides an important measure of the nature of mathematical problems and how they are solved during a lesson.

Mathematical processes suggested by problem statements

The problem implementation analysis group classified the statements of mathematical problems as one of three types based on the kind of mathematical processes implied by the statements: using procedures, stating concepts, and making connections. Because some public interaction was needed to examine the way in which the same problems were solved, this analysis was applied to all independent and concurrent problems for which a solution was reached publicly. Switzerland was not included in this analysis because English transcripts were not available for all lessons as some of the coding was conducted in Switzerland (Jacobs et al. forthcoming).

The three types of problem statements were defined as follows:

- *Using procedures:* Problem statements that suggested the problem was typically solved by applying a procedure or set of procedures. These include arithmetic with whole numbers, fractions, and decimals, manipulating algebraic symbols to simplify expressions and solve equations, finding areas and perimeters of simple plane figures, and so on. Problem statements such as "Solve for x in the equation $2x + 5 = 6 - x$" were classified as using procedures.

- *Stating concepts:* Problem statements that called for a mathematical convention or an example of a mathematical concept. Problem statements such as "Plot the point (3, 2) on a coordinate plane" or "Draw an isosceles right triangle" were classified as stating concepts.

- *Making connections:* Problem statements that implied the problem would focus on constructing relationships among mathematical ideas, facts, or procedures. Often, the problem statement suggested that students would engage in special forms of mathematical reasoning such as conjecturing, generalizing, and verifying. Problem statements such as "Graph the equations $y = 2x + 3$, $2y = x - 2$, and $y = -4x$, and examine the role played by the numbers in determining the position and slope of the associated lines" were classified as making connections.

Figure 5.8 shows that in all the countries, except Japan, at least 57 percent of the problem statements per eighth-grade mathematics lesson focused on using procedures. Hong Kong SAR lessons contained a larger percentage of problem statements classified as using procedures (84 percent) than all the other countries except the Czech Republic (77 percent). Problem statements that focused on stating concepts were found in Australian lessons (24 percent) more frequently than in the Czech, Hong Kong SAR, and Japanese lessons (which ranged from 4 percent to 7 percent). Although mathematics lessons in all the countries included problem statements that focused on making connections, the lessons from Japan contained a larger percentage of these problems (54 percent) than all the other countries except the Netherlands (24 percent).

Chapter 5
Instructional Practices: How Mathematics Was Worked On

FIGURE 5.8. Average percentage of problems per eighth-grade mathematics lesson of each problem statement type, by country: 1999

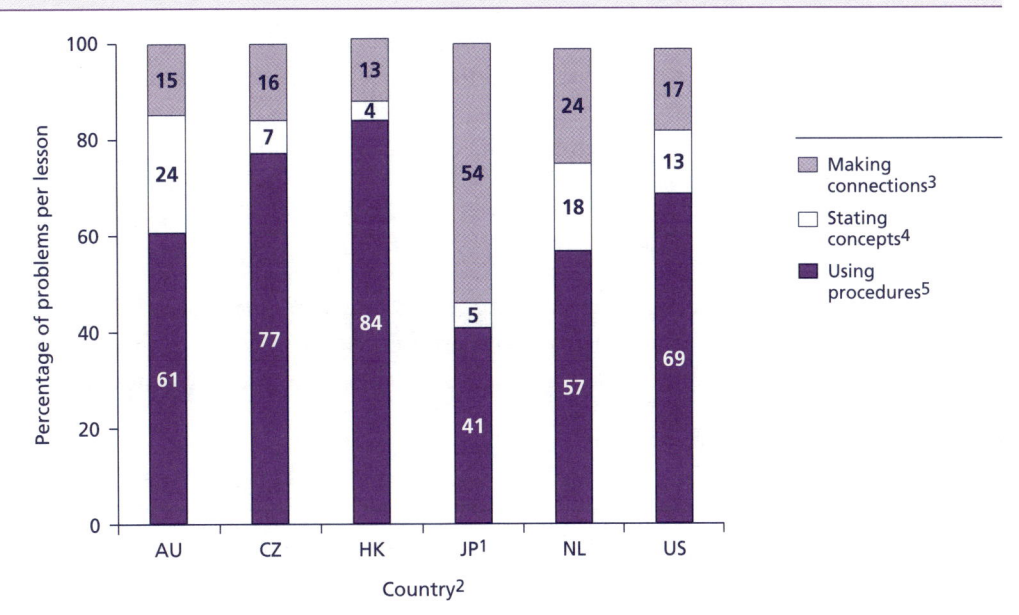

Country	Using procedures	Stating concepts	Making connections
AU	61	24	15
CZ	77	7	16
HK	84	4	13
JP[1]	41	5	54
NL	57	18	24
US	69	13	17

[1] Japanese mathematics data were collected in 1995.
[2] AU=Australia; CZ=Czech Republic; HK=Hong Kong SAR; JP=Japan; NL=Netherlands; and US=United States.
[3] Making connections: JP>AU, CZ, HK, US.
[4] Stating concepts: AU>CZ, HK, JP; NL, US>HK, JP.
[5] Using procedures: CZ>JP, NL; HK>AU, JP, NL, US; US>JP.

NOTE: Analyses do not include answered-only problems (i.e., problems that were completed prior to the videotaped lesson and only their answers were shared). For each country, average percentage was calculated as the sum of the percentage within each lesson, divided by the number of lessons. English transcriptions of Swiss lessons were not available for mathematical processes analyses. Percentages may not sum to 100 because of rounding. The tests for significance take into account the standard error for the reported differences. Thus, a difference between averages of two countries may be significant while the same difference between two other countries may not be significant.
SOURCE: U.S. Department of Education, National Center for Education Statistics, Third International Mathematics and Science Study (TIMSS), Video Study, 1999.

Using the same information in another way, an examination within each country of the relative emphases of the types of problems per lesson implied by the problem statements shows that in five of the six countries where data are available, a greater percentage of problems per lesson were presented as using procedures than either making connections or stating concepts. The exception to this pattern was Japan, where there was no detectable difference in the percentage of problems per lesson that were presented as using procedures compared to those presented as making connections.

Mathematical processes used when solving problems

A key aspect of this analysis involved following each problem from its introduction through the problem statement to its conclusion as the solution was stated publicly. A key question was whether the same kinds of mathematical processes implied by the problem statement were made explicit when solving the problem or whether the nature of the processes changed as the problem was being solved and discussed publicly.

Categories of mathematical processes for solving problems were the three types of processes defined for problem statements plus an additional category—giving results only:

- *Giving results only:* The public work consisted solely of stating an answer to the problem without any discussion of how or why it was attained.

- *Using procedures:* The problem was completed algorithmically, with the discussion focusing on steps and rules rather than underlying mathematical concepts.

- *Stating concepts:* Mathematical properties or definitions were identified while solving the problem, with no discussion about mathematical relationships or reasoning. This included, for example, stating the name of a property as the justification for a response, but not stating why this property would be appropriate for the current situation.

- *Making connections:* Explicit references were made to the mathematical relationships and/or mathematical reasoning involved while solving the problem.

Each problem was classified into exactly one of the four categories based on the mathematical processes that were made explicit during the problem solving phase. This phase began after the problem was stated and lasted until the discussion about the problem ended.

Figure 5.9 shows that from 33 to 36 percent of problems per eighth-grade mathematics lesson, on average, in Australia, the Czech Republic, and the United States, were completed publicly by giving results only. These were larger percentages than those found in the other three countries (Switzerland was not included in this analysis). Giving results only occurred least frequently in Japanese lessons. Using procedures ranged from 27 to 55 percent of problems per lesson across all the countries, and was found for a higher percentage of problems in the United States than in the Czech Republic, Japan, and the Netherlands. From 8 to 33 percent of problems per lesson were solved and discussed publicly by stating concepts, with the smallest percentage occurring in the United States. A higher percentage of problems per lesson were solved publicly through making connections in Japanese lessons (37 percent) than in all the countries except the Netherlands (22 percent). Australian and U.S. lessons contained the smallest percentages of problems implemented as making connections (2 percent and 1 percent of problems per lesson, respectively).

FIGURE 5.9. Average percentage of problems per eighth-grade mathematics lesson solved by explicitly using processes of each type, by country: 1999

Country	Giving results only[6]	Using procedures[5]	Stating concepts[4]	Making connections[3]
AU	36	41	20	2
CZ	33	38	19	10
HK	15	48	24	12
JP[1]	3	27 / 33	37	—
NL	11	36 / 32	22	—
US	36	55	8	1

[1]Japanese mathematics data were collected in 1995.
[2]AU=Australia; CZ=Czech Republic; HK=Hong Kong SAR; JP=Japan; NL=Netherlands; and US=United States.
[3]Making connections: CZ, HK, NL>AU, US; JP>AU, CZ, HK, US.
[4]Stating concepts: AU, CZ, HK, JP>US; NL>CZ, US.
[5]Using procedures: HK>JP; US>CZ, JP, NL.
[6]Giving results only: AU, CZ, US>HK, JP, NL; HK, NL>JP.

NOTE: Analyses only include problems with a publicly presented solution. Analyses do not include answered-only problems (i.e., problems that were completed prior to the videotaped lesson and only their answers were shared). For each country, average percentage was calculated as the sum of the percentage within each lesson, divided by the number of lessons. English transcriptions of Swiss lessons were not available for mathematical processes analyses. Percentages may not sum to 100 because of rounding.

SOURCE: U.S. Department of Education, National Center for Education Statistics, Third International Mathematics and Science Study (TIMSS), Video Study, 1999.

The results presented in figures 5.8 and 5.9 suggest that the processes made explicit for students while solving problems were not necessarily identical to those suggested by the problem statements. By tracing each problem through the lesson, it is possible to see what happened to problems of various types as they were being solved. It is possible, for example, to see whether a problem that began as using procedures, based on the problem statement, was retained as a using procedures problem as it was solved or whether it was transformed into a problem in which other kinds of processes were made visible. The next set of analyses examines the relationship between the way in which a problem was begun and the way in which it was solved.

Mathematical processes used when solving "using procedures" problems

By definition, problem statements classified as using procedures implied that such problems would likely be solved by applying standard procedures without examining the underlying mathematical concepts [Video clip example 5.9]. As shown in figure 5.10, between 42 and 65 percent of problems with using procedures problem statements retained a using procedures implementation per eighth-grade mathematics lesson in each country. In some cases, during the implementation of the problem, mathematical concepts and relationships were made explicit and publicly discussed. Such problems accounted for, on average, 9 to 22 percent of the using procedures problems per lesson in Hong Kong SAR, Japan, and the Netherlands—all larger percentages than in Australia.

Teaching Mathematics in Seven Countries
Results From the TIMSS 1999 Video Study

FIGURE 5.10. Average percentage of using procedures problems per eighth-grade mathematics lesson solved by explicitly using processes of each type, by country: 1999

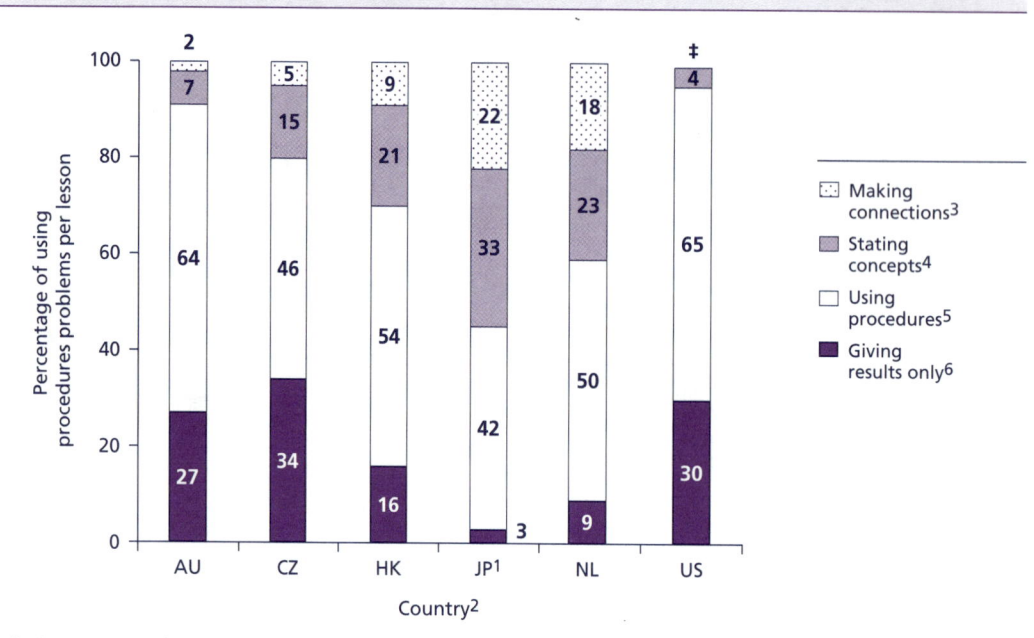

‡Reporting standards not met. Too few cases to be reported.
[1]Japanese mathematics data were collected in 1995.
[2]AU=Australia; CZ=Czech Republic; HK=Hong Kong SAR; JP=Japan; NL=Netherlands; and US=United States.
[3]Making connections: HK, JP, NL>AU; JP, NL>CZ.
[4]Stating concepts: HK, JP, NL>AU, US; CZ>US.
[5]Using procedures: US>CZ, JP.
[6]Giving results only: AU>JP; CZ>HK, JP, NL; HK>JP; US>HK, JP, NL.
NOTE: Analyses only include problems with a publicly presented solution. Analyses do not include answered-only problems (i.e., problems that were completed prior to the videotaped lesson and only their answers were shared). For each country, average percentage was calculated as the sum of the percentage within each lesson, divided by the number of lessons. English transcriptions of Swiss lessons were not available for mathematical processes analyses. Percentages may not sum to 100 because of rounding and data not reported. Lessons with no using procedures problem statements were excluded from these analyses.
SOURCE: U.S. Department of Education, National Center for Education Statistics, Third International Mathematics and Science Study (TIMSS), Video Study, 1999.

Mathematical processes used when solving "stating concepts" problems

As defined earlier, stating concepts problem statements asked students to provide a mathematical convention or an example of a mathematical concept. Due to the nature of these problems, the problem solving phase was classified as one of only three types: giving results only, stating concepts [Video clip example 5.10], or making connections [Video clip example 5.11]. Suppose, for example, the problem statement was "draw an isosceles right triangle." By definition, the problem solving phase could not include processes classified as using procedures, but it could become a giving results only problem (if the public presentation involved simply drawing a triangle that looked like it had a right angle and two congruent sides, without mentioning these facts or identifying them on the drawing), it could retain its stating concepts character (if the mathematical properties of the triangle were identified but not examined or discussed in detail), or it could become a making connections problem (if related mathematical ideas, concepts, or facts were examined—perhaps in a discussion of why certain properties must hold true for all right isosceles triangles).

Figure 5.11 shows that 100 percent of stating concepts problems in Japan were solved as stating concepts problems, a higher percentage than in the other countries (as stated earlier, Switzerland was not included in this analysis). In the United States, 61 percent of stating concepts problems per eighth-grade mathematics lesson, on average, were solved by giving a result only, a higher percentage than in Hong Kong SAR and the Netherlands. Where reliable estimates could be calculated, the percentage of stating concepts problems solved by making connections ranged from 2 to 13 percent.

FIGURE 5.11. Average percentage of stating concepts problems per eighth-grade mathematics lesson solved by explicitly using processes of each type, by country: 1999

Country	Giving results only	Stating concepts	Making connections
AU	45	54	2
CZ	48	38	13
HK	22	78	‡
JP[1]	‡	100	‡
NL	30	62	8
US	61	28	11

‡Reporting standards not met. Too few cases to be reported.
[1]Japanese mathematics data were collected in 1995.
[2]AU=Australia; CZ=Czech Republic; HK=Hong Kong SAR; JP=Japan; NL=Netherlands; and US=United States.
[3]Making connections: No differences detected.
[4]Stating concepts: JP>AU, CZ, HK, NL, US; HK>CZ, US ; NL>US.
[5]Giving results only: US>HK, NL.
NOTE: Analyses only include problems with a publicly presented solution. Analyses do not include answered-only problems (i.e., problems that were completed prior to the videotaped lesson and only their answers were shared). Lessons with no stating concepts problem statements were excluded from these analyses. For each country, average percentage was calculated as the sum of the percentage within each lesson, divided by the number of lessons. English transcriptions of Swiss lessons were not available for mathematical processes analyses. Percentages may not sum to 100 because of rounding and data not reported.
SOURCE: U.S. Department of Education, National Center for Education Statistics, Third International Mathematics and Science Study (TIMSS), Video Study, 1999.

Mathematical processes used when solving "making connections" problems

As defined earlier, problem statements classified as making connections were those that implied students would be constructing relationships among mathematical ideas, facts, or procedures, and possibly engaging in special forms of mathematical reasoning such as conjecturing, generalizing, and verifying. Figure 5.12 shows that such mathematical processes were not always made visible when these problems were solved during the lesson [Video clip example 5.12]. In Australia and the United States, 8 percent and less than 1 percent,[2] respectively, of making connections problems were solved by making connections. These percentages were smaller than in the other countries, which ranged on average from 37 to 52 percent (as stated earlier, Switzerland was not included in this analysis). Instead of solving these problems publicly through making

---

[2]The percentage of cases in the United States rounded to zero.

connections, teachers and students in Australia and the United States often solved them by giving results only (38 percent and 33 percent, respectively) or, in the United States, by using procedures (59 percent).

FIGURE 5.12. Average percentage of making connections problems per eighth-grade mathematics lesson solved by explicitly using processes of each type, by country: 1999

Country	Giving results only	Using procedures	Stating concepts	Making connections
AU	38	31	23	8
CZ	8	16	24	52
HK	5	18	31	46
JP[1]	3	20	29	48
NL	4	19	40	37
US	33	59	8	#

#Rounds to zero.
[1]Japanese mathematics data were collected in 1995.
[2]AU=Australia; CZ=Czech Republic; HK=Hong Kong SAR; JP=Japan; NL=Netherlands; and US=United States.
[3]Making connections: CZ, HK, JP, NL>AU, US.
[4]Stating concepts: JP, NL>US.
[5]Using procedures: US>CZ, HK, JP, NL.
[6]Giving results only: AU, US>CZ, HK, JP, NL.
NOTE: Analyses only include problems with a publicly presented solution. Analyses do not include answered-only problems (i.e., problems that were completed prior to the videotaped lesson and only their answers were shared). Lessons with no making connections problem statements were excluded from these analyses. For each country, average percentage was calculated as the sum of the percentage within each lesson, divided by the number of lessons. English transcriptions of Swiss lessons were not available for mathematical processes analyses. Percentages may not sum to 100 because of rounding. The tests for significance take into account the standard error for the reported differences. Thus, a difference between averages of two countries may be significant while the same difference between two other countries may not be significant.
SOURCE: U.S. Department of Education, National Center for Education Statistics, Third International Mathematics and Science Study (TIMSS), Video Study, 1999.

Private Work Assignments

To this point, most of the findings presented in this chapter apply to events that occurred publicly during the lesson, that were visible for all students to see or hear. But, in every country some lesson time, on average, was devoted to private work (chapter 3, table 3.6). That is, eighth-grade students were asked to complete mathematical problems by working on their own or in small groups. Less information was available to evaluate the mathematical processes in which students engaged during private time than public time, but it was possible to classify students' private work into one of two categories: (1) repeating procedures that had been demonstrated earlier in the lesson or learned in previous lessons, or (2) doing something other than repeating learned procedures. "Something other" might have been developing solution procedures that were new for the students or modifying solution procedures they already had learned.

Each private work segment was marked for whether students worked on an assignment with problems that required them to repeat procedures [Video clip example 5.13], do something other than repetition [Video clip example 5.14], or do a mix of repetition and something other than repetition. An assignment was considered mixed when it contained several problems, at least one of which required repetition and at least one of which required something other than repetition.

Figure 5.13 shows that in Japan, on average, a smaller percentage of private work time per eighth-grade mathematics lesson was spent repeating procedures (28 percent) compared to all the other countries. The percentages in other countries ranged from 62 to 84. It follows then that in Japan, on average, a larger percentage of private work time per lesson was devoted to doing something other than repeating procedures or doing a mix of repeating and something other than repeating (65 percent) compared to all the other countries (ranging from 9 to 25 percent).

FIGURE 5.13. Average percentage of private work time per lesson devoted to repeating procedures and something other than repeating procedures or mix, by country: 1999

[1] Japanese mathematics data were collected in 1995.
[2] AU=Australia; CZ=Czech Republic; HK=Hong Kong SAR; JP=Japan; NL=Netherlands; SW=Switzerland; and US=United States.
[3] Other than repeating procedures or mix: JP>AU, CZ, HK, NL, SW, US; AU, SW>US.
[4] Repeating procedures: AU, NL, SW, US>JP; CZ, HK>JP, SW.
NOTE: For each country, average percentage was calculated as the sum of the percentage within each lesson, divided by the number of lessons. Percentages may not sum to 100 because some private work segments were marked as "not able to make judgment."
SOURCE: U.S. Department of Education, National Center for Education Statistics, Third International Mathematics and Science Study (TIMSS), Video Study, 1999.

Differences among other countries in how students spent their private work time also were found (figure 5.13). In Switzerland, a smaller percentage of private work time per eighth-grade mathematics lesson was devoted to assignments involving repeating procedures (62 percent) than in either the Czech Republic or Hong Kong SAR (84 percent and 81 percent, respectively). Conversely, more Australian and Swiss private work time per lesson was devoted to doing something other than repeating procedures (24 percent and 25 percent, respectively) than was the case in the United States (9 percent).

The Nature of Non-Problem Segments

Although the majority of time in the sampled lessons was spent working on problems (see chapter 3, figure 3.3), eighth-grade students also had opportunities to learn mathematics during non-problem segments (i.e., while working on mathematics outside the context of a problem). These learning opportunities depended on the type of information provided or the kind of activity in which the class engaged. Each non-problem segment in the videotaped lessons was coded into at least one of four, non-mutually exclusive categories:

- *Mathematical information:* Presenting or discussing new material or material previously presented, perhaps through a brief lecture by the teacher [Video clip example 5.15].

- *Contextual information:* Describing the goal for the lesson, presenting historical background, introducing a real-life example, or relating mathematical ideas discussed in the current lesson to a past or future lesson [Video clip example 5.16].

- *Mathematical activity:* Playing games (e.g., bingo or hangman), or completing other tasks that were not mathematical problems (e.g., a worksheet page that contained a word search for mathematical terms).

- *Announcements:* Announcing a homework assignment or test, or clarifying an assignment (but not discussing mathematical information that was on the test or assignment) [Video clip example 5.17].

Announcements contained no substantive mathematical information and, consequently, would seem to provide little opportunity, by themselves, for student learning. Using a similar logic, mathematical information and contextual information might provide the most direct opportunities for students' mathematics learning. Table 5.5 shows the percentage of non-problem segments per eighth-grade mathematics lesson that were judged to contain each type of activity or information.

TABLE 5.5. Average percentage of non-problem segments per lesson coded as mathematical information, contextual information, mathematical activity, and announcements, by country: 1999

Country	Mathematical information[2]	Contextual information[3]	Mathematical activity[4]	Announcements[5]
Australia (AU)	67	56	6	33
Czech Republic (CZ)	54	63	4	22
Hong Kong SAR (HK)	73	53	2	28
Japan[1] (JP)	83	67	7	11
Netherlands (NL)	69	47	‡	23
Switzerland (SW)	54	59	3	29
United States (US)	73	49	10	21

‡Reporting standards not met. Too few cases to be reported.
[1]Japanese mathematics data were collected in 1995.
[2]Mathematical information: HK>CZ, SW; JP>AU, CZ, SW; US>CZ.
[3]Contextual information: JP>US.
[4]Mathematical activity: No differences detected.
[5]Announcements: AU, HK, SW>JP.

NOTE: For each country, average percentage was calculated as the sum of the percentage within each lesson, divided by the number of lessons. Percentage may not sum to 100 because segments could be coded as more than one type. The tests for significance take into account the standard error for the reported differences. Thus, a difference between averages of two countries may be significant while the same difference between two other countries may not be significant.
SOURCE: U.S. Department of Education, National Center for Education Statistics, Third International Mathematics and Science Study (TIMSS), Video Study, 1999.

Across all the countries, at least 54 percent of the non-problem segments per eighth-grade mathematics lesson, on average, contained mathematical information and at least 47 percent contained contextual information. Japanese teachers used non-problem segments to present mathematical information more often than teachers in Australia, the Czech Republic, and Switzerland.

No differences among the countries were found for the percentage of non-problem segments devoted to mathematical activity (between 2 and 10 percent). Japanese teachers were less likely to use non-problem segments for announcements (11 percent of non-problem segments per lesson) than were teachers in Australia, Hong Kong SAR, and Switzerland (33, 28, and 29 percent, respectively).

Opportunities to Talk

An enduring controversy in teaching research is the contribution of active student participation in classroom discourse (Goldenberg 1992/1993). Although most studies show that teachers talk the majority of the time while their students are listeners (Goodlad 1984; Hiebert and Wearne 1993; Hoetker and Ahlbrand 1969; Tharp and Gallimore 1989), there is disagreement over whether this pattern adversely affects learning. Some argue that limited student talk reduces learning opportunities to those excessively weighted toward low-level skills and factually oriented instruction (Bunyi 1997; Cazden 1988; Knapp and Shields 1990). Advocates of student talk also suggest that student interaction increases the opportunities for students to elaborate, clarify, and reorganize their own thinking (Ball 1993; Hatano 1988). Others argue that student learning is best fostered by explicit or direct teaching, such as stating an objective and providing step-by-step instruction—which necessarily awards teachers substantially more talk opportunities than students (e.g., Gage 1978; Rosenshine and Stevens 1986; Walberg 1990). A third view suggests that the optimum ratio of teacher to student talk is a function of the content students are to learn (Goldenberg 1992/1993). In sum, there is no broad consensus regarding the impact of a larger role for students in classroom discourse.

Classroom discourse research suggests that students must utter more than single words or short phrases before their participation can qualify as active or be indicative of opportunities for extended discussion of academic content (Cazden 1988). Word-based measures provide a proxy indication of whether that is the case, and to what extent classroom discourse is teacher-dominated in terms of opportunities to talk.

Computer-assisted analyses were applied in the TIMSS 1999 Video Study to English transcripts of eighth-grade mathematics lessons. In the case of the Czech Republic, Japan, and the Netherlands, all lessons were translated from the respective native languages, while in the case of Hong Kong SAR, 34 percent of the lessons were conducted in English, and 66 percent needed to be translated. English translations of all Swiss lessons were not available so Switzerland was not included in these analyses. Analyses based on same-language transcripts makes the speech across countries more comparable. Transcriber/translators were fluent in both English and the language of the country whose lessons they translated, educated at least through eighth grade in the country whose lessons they translated, and had completed two weeks of training. A glossary was developed to standardize translation of special terms within each country.

Computer-assisted text analyses were applied by the text analysis group (see appendix A) to all segments of public interaction[3] to quantify how often eighth-grade students talked during mathematics lessons.

Before presenting the results, it is important to note that student talk was recorded and transcribed from a microphone worn by the teacher as well as one mounted on each of the two video cameras. When many students spoke at once or made remarks out of range of the microphones, transcribers were sometimes able to detect that something was said without making out the words. The percent of inaudible student utterances ranged from 5 to 10 percent in all the countries, with one exception: eighth-grade mathematics lessons in the Netherlands had significantly more inaudible student utterances than the other countries (27 percent).[4]

A first indicator of how talk was shared between teachers and students is the total number of words spoken by teachers and students during public interaction. Because countries differed in the average duration of their lessons (see chapter 3, table 3.1) and in the average amount of public interaction time (see chapter 3, table 3.6), these comparisons examined teacher and student talk standardized for 50 minutes of lesson time. As shown in figure 5.14, Hong Kong SAR and U.S. eighth-grade mathematics teachers talked more than their counterparts in Japan, but otherwise there were no differences detected among countries in the amount of teacher talk adjusted per 50 minutes of public interaction. Hong Kong SAR students spoke significantly fewer words compared to their peers in the Netherlands and United States. In all the countries, teachers spoke more words than did students per lesson.

---

[3]Public interaction was defined as a public presentation by the teacher or one or more students intended for all students.
[4]Post hoc comparisons on the number of inaudible student utterances yielded the following statistically significant differences: CZ>HK, JP; NL>AU, CZ, HK, JP, US.

FIGURE 5.14. Average number of teacher and student words per eighth-grade mathematics lesson, by country: 1999

[Bar chart showing average number of teacher words and student words by country:
- AU: Teacher 5536, Student 810
- CZ: Teacher 5452, Student 824
- HK: Teacher 5798, Student 640
- JP[1]: Teacher 5148, Student 766
- NL: Teacher 5360, Student 1016
- US: Teacher 5902, Student 1018]

[1] Japanese mathematics data were collected in 1995.
[2] AU=Australia; CZ=Czech Republic; HK=Hong Kong SAR; JP=Japan; NL=Netherlands; and US=United States.
[3] Average number of teacher words per 50 minutes of public interaction: HK, US>JP.
[4] Average number of student words per 50 minutes of public interaction: NL, US>HK.

NOTE: Analyses based on English transcripts. English transcriptions of Swiss lessons were not available for text analyses. For each lesson, the number of words spoken by the teacher or the students was standardized for 50 minutes of public interaction using the following formula: number of words divided by the amount of public interaction time multiplied by 50.

SOURCE: U.S. Department of Education, National Center for Education Statistics, Third International Mathematics and Science Study (TIMSS), Video Study, 1999.

A second indicator of the relative share of talk time afforded students during public interaction is the ratio of teacher to student talk—a metric less sensitive to possible effects of using English translations. As displayed in figure 5.15, Hong Kong SAR eighth-grade mathematics teachers spoke significantly more words relative to their students (16:1) than did teachers in Australia (9:1), the Czech Republic (9:1), and the United States (8:1).

| FIGURE 5.15. | Average number of teacher words to every one student word per eighth-grade mathematics lesson, by country: 1999 |

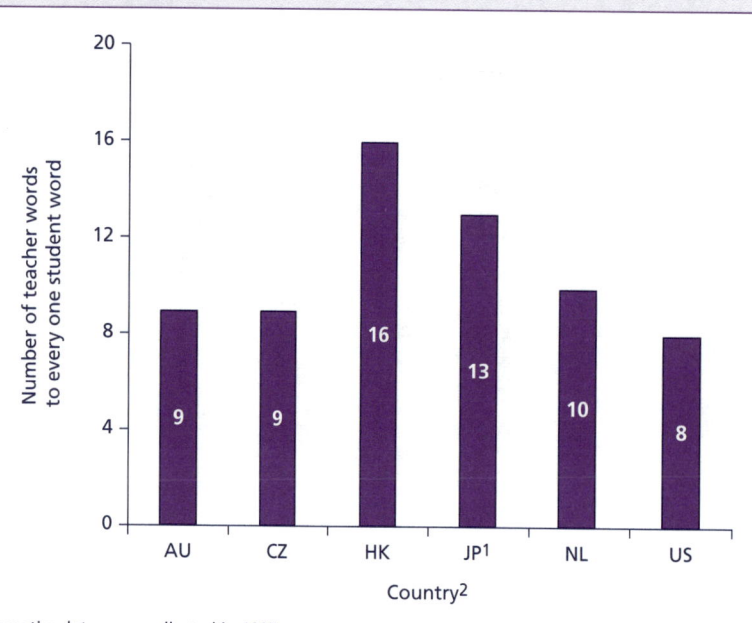

Bar chart — Number of teacher words to every one student word:
- AU: 9
- CZ: 9
- HK: 16
- JP[1]: 13
- NL: 10
- US: 8

[1] Japanese mathematics data were collected in 1995.
[2] AU=Australia; CZ=Czech Republic; HK=Hong Kong SAR; JP=Japan; NL=Netherlands; and US=United States.
NOTE: HK>AU, CZ, US. Analyses based on English transcripts. English transcriptions of Swiss lessons were not available for text analyses.
SOURCE: U.S. Department of Education, National Center for Education Statistics, Third International Mathematics and Science Study (TIMSS), Video Study, 1999.

A third indicator of opportunity to talk during lessons is the length of each utterance. For purposes of this analysis, an utterance was defined as talk by one speaker uninterrupted by another speaker. Overlapping speech was transcribed with each speaker's contribution recorded as a separate utterance, if audible. Transcribers were instructed to identify a new utterance any time a new speaker began talking, and to note who was speaking (e.g., teacher or student). Longer student utterances are often interpreted as indicators of opportunities for fuller student participation in classroom discussions, whereas short utterances often reflect faster-paced "back and forth" exchanges between teachers and students. In faster-paced exchanges, students are typically restricted to single words or short phrases (Cazden 1988; Goldenberg 1992/1993; Tharp and Gallimore 1989).

As can be deduced from figure 5.16, between 71 and 82 percent of all teacher utterances on average per lesson contained more than 5 words. In contrast, between 23 and 34 percent of student utterances on average per lesson contained more than 5 words (figure 5.17). In none of these countries did the number of longer student utterances (10+ words) exceed 9 percent. However, as shown in figures 5.16 and 5.17, there were differences between countries on specific dimensions indicating that lessons in some countries provided different opportunities than others, although in absolute terms none of these differences are large.

Chapter 5
Instructional Practices: How Mathematics Was Worked On
111

FIGURE 5.16. Average percentage of teacher utterances of each length per eighth-grade mathematics lesson, by country: 1999

Country	1–4 word teacher utterances[4]	5+ word teacher utterances[5]	25+ word teacher utterances[3,6]
AU	21	79	35
CZ	20	80	35
HK	18	82	41
JP[1]	29	71	38
NL	24	77	25
US	19	81	33

[1] Japanese mathematics data were collected in 1995.
[2] AU=Australia; CZ=Czech Republic; HK=Hong Kong SAR; JP=Japan; NL=Netherlands; and US=United States.
[3] Percentage of teacher utterances that were 25+ words is a subset of teacher utterances that were 5+ words.
[4] Percent of teacher utterances that were 1–4 words: JP>AU, CZ, HK, US; NL>CZ, HK, US.
[5] Percent of teacher utterances that were 5+ words: AU, CZ, HK, US>JP; HK>NL.
[6] Percent of teacher utterances that were 25+ words: AU, CZ, JP, US>NL; HK>AU, CZ, NL, US.

NOTE: Analyses based on English transcripts. English transcriptions of Swiss lessons were not available for text analyses. For each country, average percentage was calculated as the sum of the percentage within each lesson, divided by the number of lessons. Percentages of 1–4 word teacher utterances and 5+ word teacher utterances may not sum to 100 because of rounding.
SOURCE: U.S. Department of Education, National Center for Education Statistics, Third International Mathematics and Science Study (TIMSS), Video Study, 1999.

FIGURE 5.17. Average percentage of student utterances of each length per eighth-grade mathematics lesson, by country: 1999

Country	1–4 word	5+ word	10+ word
AU	71	29	7
CZ	66	34	9
HK	77	23	4
JP[1]	70	30	9
NL	72	29	7
US	71	30	7

[1] Japanese mathematics data were collected in 1995.
[2] AU=Australia; CZ=Czech Republic; HK=Hong Kong SAR; JP=Japan; NL=Netherlands; and US=United States.
[3] Percentage of student utterances that were 10+ words is a subset of student utterances that were 5+ words.
[4] Percent of student utterances that were 1–4 words: HK>AU, CZ, JP, NL, US; NL, US>CZ.
[5] Percent of student utterances that were 5+ words: AU, JP, NL, US>HK; CZ>HK, NL, US.
[6] Percent of student utterances that were 10+ words: AU, CZ, JP, NL, US>HK.

NOTE: Analyses based on English transcripts. English transcriptions of Swiss lessons were not available for text analyses. For each country, average percentage was calculated as the sum of the percentage within each lesson, divided by the number of lessons. Percentages of 1–4 word student utterances and 5+ word student utterances may not sum to 100 because of rounding. The tests for significance take into account the standard error for the reported differences. Thus, a difference between averages of two countries may be significant while the same difference between two other countries may not be significant.

SOURCE: U.S. Department of Education, National Center for Education Statistics, Third International Mathematics and Science Study (TIMSS), Video Study, 1999.

Eighth-grade mathematics lessons taught by Japanese teachers had significantly more short utterances by teachers (1, 2, 3, or 4 words in length) and significantly fewer 5+ word utterances by teachers than those in Australia, the Czech Republic, Hong Kong SAR, and the United States. As well, teachers in the Netherlands had more short utterances than did teachers in the Czech Republic, Hong Kong SAR, and the United States. Lessons taught by Dutch teachers also had fewer "mini-lectures" (utterances of 25+ words) than did lessons taught by mathematics teachers in Australia, the Czech Republic, Japan, and the United States. The mathematics lessons in Hong Kong SAR were distinct from lessons in Australia, the Czech Republic, the Netherlands, and the United States by having significantly more mini-lectures delivered by teachers.

Eighth-grade mathematics lessons in Hong Kong SAR had more short utterances (1, 2, 3, or 4 words) and fewer longer utterances (5+ and 10+ words) delivered by students than lessons in all the other countries for which these analyses were conducted. Lessons in the Czech Republic had significantly more 5+ word utterances delivered by eighth-graders than lessons in Hong Kong SAR, the Netherlands, and the United States, and fewer short utterances (1, 2, 3, or 4 words) delivered by students than lessons in the Netherlands and the United States.

In broad terms, the eighth-grade mathematics lessons in all the countries revealed many brief opportunities for students to talk while mathematical work was being done, and very few long opportunities. This is similar to the pattern often reported in the literature, in which teachers talk and students listen (Cazden 1988; Goodlad 1984; Hiebert and Wearne 1993; Hoetker and Ahlbrand 1969; Tharp and Gallimore 1989).

Resources Used During the Lesson

This chapter concludes by identifying the kinds of supportive materials that were used during the videotaped eighth-grade mathematics lessons. These include audio-visual equipment, print material, particular types of physical materials, and technology. In all cases, the resource was marked if it was used at any point in the lesson. If the materials and technology were present but not used, they were not included in these analyses.

- *Chalkboard:* Included chalkboards and whiteboards.

- *Projector:* Included overhead, video, and computer projectors.

- *Textbook/worksheets:* Included textbooks, review sheets, study sheets, worksheets, and the like.

- *Special mathematics materials:* Included materials such as graph paper, graph boards, hundreds tables, geometric solids, base-ten blocks, rulers, measuring tape, compasses, protractors, and computer software that simulates constructions of models [Video clip example 5.18]. When special mathematics materials were used for presenting or solving a mathematics problem, they were included also in the earlier description of physical materials and analyzed by percentage of problems, figure 5.3.

- *Real-world objects:* Included objects such as cans, beans, toothpicks, maps, dice, newspapers, magazines, and springs [Video clip example 5.19]. When real-world objects were used for presenting or solving a mathematics problem, they were included also in the earlier description of physical materials and analyzed by percentage of problems, figure 5.3.

- *Calculators:* Included computational and graphing calculators, but each was marked separately [Video clip example 5.20].

- *Computers* [Video clip example 5.21].

Table 5.6 depicts the percentage of eighth-grade mathematics lessons during which a chalkboard, projector, textbook or worksheet, special mathematical materials, and real-world objects were used. Chalkboards were used in a smaller percentage of lessons in the United States (71 percent) than in all the other countries except Switzerland. Perhaps as a substitute, projectors were used in a higher percentage of lessons in Switzerland and the United States (49 percent and 59 percent, respectively) than they were in all the other countries (ranging from 3 to 23 percent). Nearly all lessons in all the countries used either a textbook or worksheet, ranging from 91 percent of Australian lessons to 100 percent of Czech and Dutch lessons.

| TABLE 5.6. | Percentage of eighth-grade mathematics lessons during which a chalkboard, projector, textbook/worksheet, special mathematics material, and real-world object were used, by country: 1999 |

	Resources used				
Country	Chalkboard[2]	Projector[3]	Textbook/ worksheet[4]	Special mathematics materials[5]	Real-world objects[6]
	Percent				
Australia (AU)	97	16	91	44	21
Czech Republic (CZ)	100	23	100	66	10
Hong Kong SAR (HK)	97	12	99	30	4
Japan[1] (JP)	98	11	92	86	19
Netherlands (NL)	96	3	100	81	7
Switzerland (SW)	90	49	95	32	20
United States (US)	71	59	98	44	15

[1] Japanese mathematics data were collected in 1995.
[2] Chalkboard: AU, CZ, HK, JP, NL>US.
[3] Projector: CZ>NL; SW, US>AU, CZ, HK, JP, NL.
[4] Textbook/worksheet: HK>JP.
[5] Special mathematics materials: CZ>HK, SW; JP>AU, CZ, HK, SW, US; NL>AU, HK, SW, US.
[6] Real-world objects: AU, SW, US>HK.

NOTE: Percentage of lessons reported for Japan with respect to chalkboard, projector, and textbook/worksheet use differs from that reported in Stigler et al. (1999) because the definitions were changed for the current study. The tests for significance take into account the standard error for the reported differences. Thus, a difference between averages of two countries may be significant while the same difference between two other countries may not be significant.

SOURCE: U.S. Department of Education, National Center for Education Statistics, Third International Mathematics and Science Study (TIMSS), Video Study, 1999.

Special mathematics materials were used in 86 percent of Japanese eighth-grade mathematics lessons, a higher percentage than Australia, the Czech Republic, Hong Kong SAR, Switzerland, and the United States (ranging from 30 to 66 percent of lessons). This is consistent with the findings presented earlier on the use of physical materials for solving individual problems in Japan (figure 5.3). As stated earlier, however, the use of physical materials seemed to be related to the mathematical topic covered in the lesson, and the relative differences between Japan and some of the other countries changed when considering only two-dimensional geometry problems (figure 5.4).

In contrast to Japan, the Netherlands showed a different pattern when considering the percentage of lessons in which special mathematics materials were used (table 5.6, 81 percent) and the percentage of problems per lesson that involved the use of physical materials (figure 5.3, 3 percent). One possible explanation for this difference is that Dutch teachers could have used special materials for a single problem, thus accounting for the relatively high percentage of lessons but the relatively low percentage of problems. Or, special mathematics materials could have been used primarily during non-problem segments. Another explanation is that the two codes do not include the same materials—special mathematics materials included materials not counted as physical materials (e.g., graph paper).

With the increasing use of technology in all aspects of society, there is special interest in the use of calculators and computers in mathematics class (Fey and Hirsch 1992; Kaput 1992; Ruthven 1996). The use of computational calculators is a contested issue, with opponents concerned that calculator use, especially in the early grades, will limit students' computational fluency and advocates arguing that calculators are an increasingly common tool that can be used in the classroom to facilitate students' learning (National Research Council 2001a).

Figure 5.18 shows that calculators used for computation (i.e., those not equipped with graphing capability or whose graphing capability was not utilized) were found more frequently in the Netherlands (in 91 percent of lessons) than in Australia, the Czech Republic, Hong Kong SAR, Switzerland, and the United States, where the range of use was 31 percent to 56 percent of eighth-grade mathematics lessons.

FIGURE 5.18. Percentage of eighth-grade mathematics lessons during which computational calculators were used, by country: 1999

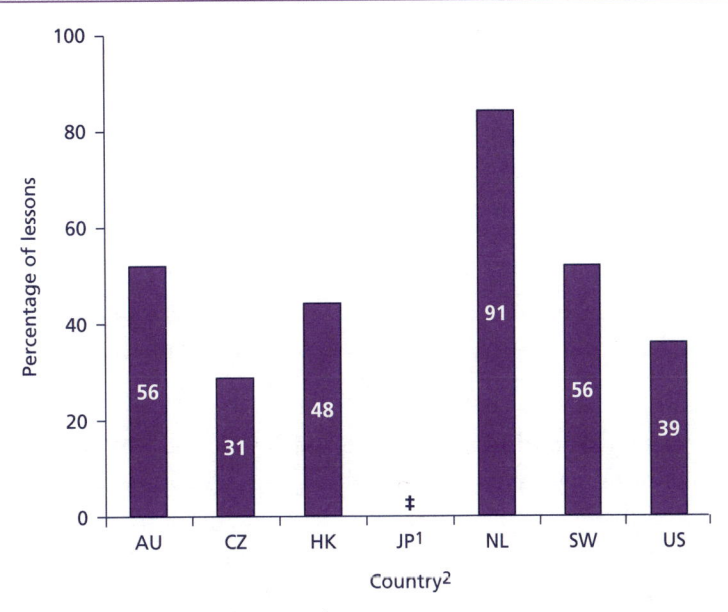

‡Reporting standards not met. Too few cases to be reported.
[1]Japanese mathematics data were collected in 1995.
[2]AU=Australia; CZ=Czech Republic; HK=Hong Kong SAR; JP=Japan; NL=Netherlands; SW=Switzerland; and US=United States.
NOTE: NL>AU, CZ, HK, SW, US; SW>CZ. The tests for significance take into account the standard error for the reported differences. Thus, a difference between averages of two countries may be significant while the same difference between two other countries may not be significant.
SOURCE: U.S. Department of Education, National Center for Education Statistics, Third International Mathematics and Science Study (TIMSS), Video Study, 1999.

Calculators used for graphing were rarely seen in the eighth-grade mathematics lessons in the participating countries, except in the United States where they were used in 6 percent of the lessons. In all the other countries, graphing calculators were observed too infrequently to calculate reliable estimates for their use.

Computers were used in 9 percent of Japanese, 5 percent of Hong Kong SAR, 4 percent of Australian, and 2 percent of Swiss eighth-grade mathematics lessons. No differences were detected in computer use among these countries (not shown in tables or figures). In the Czech Republic, the Netherlands, and the United States, computers were used too infrequently to produce reliable estimates.

Summary

In many ways, the issues addressed in this chapter are at the heart of mathematics teaching. The ways in which mathematics is presented and the ways in which teachers and students interact about the mathematics are direct indicators of the nature of teaching and the nature of learning opportunities for students. Chapters 2, 3, and 4 set the stage for this chapter, and the findings presented here provide more detailed information about eighth-grade mathematics teaching in each country.

The observations below summarize some of the key findings in this chapter. The majority of these findings have to do with the way in which mathematics problems were presented and worked on during lessons.

- Eighth-grade mathematics lessons in the Netherlands emphasized the relationships between mathematics and real-life situations to a greater extent than most of the other countries (figure 5.1). Forty-two percent of the problems per lesson in the Netherlands, on average, were set up with a real-life connection. By contrast, in the other countries, between 9 and 27 percent of the problems per lesson were set up with a real-life connection.

- Applications—problems that ask students to decide how to use procedures rather than just execute them—were a feature of eighth-grade mathematics problems in Japanese lessons (74 percent of problems per lesson) to a greater degree than in all the other countries except Switzerland (55 percent of problems per lesson) (figure 5.6).

- Eighth-grade students presenting and examining alternative solution methods for mathematics problems was not a common activity in any of the countries (tables 5.1 and 5.2), despite their discussion in the literature. Using a generous criterion, alternative solution methods were presented for 17 percent of the mathematics problems per lesson in Japan, a higher percentage than found in Australia, the Czech Republic, and Hong Kong SAR. Using stricter criteria that required the solution methods to be examined publicly and students to present at least one of the methods, the percentages dropped to 9 percent of problems per lesson in Japan and no more than 3 percent in all the other countries (table 5.3).

- Tracing the life of mathematics problems from when they were presented to when their solutions were stated publicly showed some significant differences among countries in terms of the types of problems presented and in the changes that occurred as they were solved:

  ○ A larger percentage of eighth-grade mathematics problems in Japan (54 percent) than in most of the other countries were stated in a way that suggested making mathematical connections or exploring mathematical relationships (figure 5.8). In contrast, a larger percentage of mathematics problems in Hong Kong SAR (84 percent) than in most of the other countries were stated in a way that suggested using procedures (figure 5.8).

  ○ As problems were solved, the mathematical processes that were made visible for the eighth-graders were often different from those implied by the statement of the problem (figures 5.10–5.12). For example, mathematics teachers transformed some of the problems stated as making connections into problems that focused on using procedures. Lessons taught by mathematics teachers in Australia and the United States retained the making connections focus of problems less often than lessons in the other countries (8 percent of the making connections problems and less than 1 percent of the making connections problems, respectively; see figure 5.12).

- Students in different countries were expected to do different kinds of work during private time. More private work time per lesson in the Czech Republic and Hong Kong SAR was devoted to repeating procedures students had already learned (84 and 81 percent, on average, respectively) compared to Japan and Switzerland (28 and 62 percent, on average, respectively) (figure 5.13). Eighth-grade students in Japan were expected to do something other than repeating procedures, such as adapting procedures to solve new kinds of problems, during a greater percentage of lesson time than in all the other countries (figure 5.13).

- In all the countries, eighth-grade mathematics teachers did most of the talking during the lessons. The ratio of teacher talk to student talk per lesson ranged from 8:1 (the United States) to 16:1 (Hong Kong SAR). The ratio for teacher talk to student talk was higher in lessons in Hong Kong SAR than in Australia, the Czech Republic, and the United States (figure 5.15). In broad terms, lessons taught by teachers in all the countries provided many brief opportunities for students to talk during periods of public interaction, and fewer opportunities for more extensive discussion (figure 5.17).

The findings presented in this and earlier chapters address different aspects of classroom practice. Dissecting classroom practice into separate dimensions and variables is essential for studying it, but separating teaching into distinct features can mask the fact that all of these features interact to create systems of teaching (Stigler and Hiebert 1999). It is important now to step back and ask how these findings fit together. What can be learned about teaching in each country by looking across the chapters and piecing the results together? That is the task undertaken in chapter 6.

CHAPTER 6
Similarities and Differences in Eighth-Grade Mathematics Teaching Across Seven Countries

One of the primary questions left unanswered by the TIMSS 1995 Video Study, and a question that motivated the current study, was whether teachers in all countries whose students achieve well in mathematics teach the subject in a similar way. It was tempting for some people who were familiar with the 1995 study to draw the conclusion that the method of teaching mathematics seen in the Japanese videotapes was necessary for high achievement. But in the 1995 study of teaching in Germany, Japan, and the United States, there was only one high-achieving country as measured through TIMSS—Japan. As the TIMSS 1999 Video Study includes a number of high-achieving countries, this study can shed more light on whether high-achieving countries teach in a similar manner.

The TIMSS 1999 Video Study aimed to reveal similarities and differences in teaching practices among all seven countries in the sample and to consider whether distinctive patterns of eighth-grade mathematics teaching can be detected in each country. This chapter summarizes the results presented in the earlier chapters and presents some new within-country analyses in order to address the descriptive and comparative issues that motivated this study.

For reader convenience, table 6.1 repeats the eighth-grade achievement results presented in chapter 1 for the seven participating countries on the TIMSS 1995 and the TIMSS 1999 mathematics assessments. As stated earlier, based on results from TIMSS 1995 and 1999, Hong Kong SAR and Japan were consistently high achieving relative to other countries in the TIMSS 1999 Video Study and the United States was lower achieving in mathematics than all the other countries in 1995 and lower than all but the Czech Republic of those countries that participated in 1999 (Gonzales et al. 2000).

TABLE 6.1. Average scores on the TIMSS mathematics assessment, grade 8, by country: 1995 and 1999

Country	TIMSS 1995 mathematics score[1] Average	Standard error	TIMSS 1999 mathematics score[2] Average	Standard error
Australia[3] (AU)	519	3.8	525	4.8
Czech Republic (CZ)	546	4.5	520	4.2
Hong Kong SAR (HK)	569	6.1	582	4.3
Japan (JP)	581	1.6	579	1.7
Netherlands[3] (NL)	529	6.1	540	7.1
Switzerland (SW)	534	2.7	—	—
United States (US)	492	4.7	502	4.0
International average[4]	—	—	487	0.7

—Not available.
[1]TIMSS 1995: AU>US; HK, JP>AU, NL, SW, US; JP>CZ; CZ, SW>AU, US; NL>US.
[2]TIMSS 1999: AU, NL>US; HK, JP>AU, CZ, NL, US.
[3]Nation did not meet international sampling and/or other guidelines in 1995. See Beaton et al. (1996) for details.
[4]International average: AU, CZ, HK, JP, NL, US>international average.
NOTE: Rescaled TIMSS 1995 mathematics scores are reported here (Gonzales et al. 2000). Due to rescaling of 1995 data, international average not available. Switzerland did not participate in the TIMSS 1999 assessment.
SOURCE: Gonzales, P., Calsyn, C., Jocelyn, L., Mak, K., Kastberg, D., Arafeh, S., Williams, T., and Tsen, W. (2000). *Pursuing Excellence: Comparisons of International Eighth-Grade Mathematics and Science Achievement From a U.S. Perspective, 1995 and 1999* (NCES 2001-028). U.S. Department of Education. Washington, DC: National Center for Education Statistics.

Locating Similarities and Differences in Eighth-Grade Mathematics Teaching

Two lenses can be used to provide two different views: one lens provides a wide-angle shot and considers more general features of teaching; the other lens moves in closer and focuses on the nature of the content and the way in which students and teachers interact with respect to mathematics. Each lens provides a different perspective on the data.

A Wide-Angle Lens: General Features of Teaching

From a wide-angle view, the analyses contained in this report reveal that eighth-grade mathematics lessons across the seven countries share some general features. For example, mathematics teachers in all the countries organized the average lesson to include some public whole-class work and some private individual or small-group work (table 3.6). Teachers in all the countries talked more than students, at a ratio of at least 8:1 teacher to student words (figure 5.15). At least 90 percent of lessons in all the countries used a textbook or worksheet of some kind (table 5.6). Teachers in all the countries taught mathematics largely through the solving of mathematics problems (at least 80 percent of lesson time, on average, figure 3.3). And, on average, lessons in all the countries included some review of previous content as well as some attention to new content (figure 3.8).

The suggestion that the countries share some common ways of teaching eighth-grade mathematics is consistent with an interpretation that emphasizes the similarities of teaching practices across countries because of global institutional trends (LeTendre et al. 2001). By focusing on variables that reflect general features of teaching and instructional resources, the similarities among countries in this study can be highlighted. This wide-angle lens suggests that many of the same basic ingredients were used to construct eighth-grade mathematics lessons in all of the participating countries (Anderson-Levitt 2002).

A Close-Up Lens: Mathematics Problems and How They Are Worked On

A second, close-up lens brings a different picture into focus. This closer view considers how the general features of teaching were combined and carried out during the lesson. It reveals particular differences among countries in mathematics problems and how they are worked on.

One way of examining differences among countries is to ask whether countries showed distinctive patterns of teaching. Did the mathematics lessons in one country differ from the lessons in all the other countries on particular features? This criteria sets the bar quite high, but provides one way of looking across countries for unique approaches to the teaching of mathematics. Looking across the results presented in this report, there is no country among those that participated in the study that is distinct from all the other countries on all the features examined in this study. The 1995 study seemed to point to Japan as having a more distinct way of teaching eighth-grade mathematics when compared to the other two countries in that study. Based on the results of the 1999 study, all countries exhibited some differences from the other countries on features of eighth-grade mathematics teaching. However, Japanese eighth-grade mathematics lessons differed from all the other countries in the study on 17 of the analyses related to the mathematics lessons, or 15 percent of the analyses conducted for this report.[1] Among the other countries, the Netherlands was found to be distinct from all the other participating countries on 10 analyses, or 9 percent, the next highest frequency. Among the other five countries, three differed from every other country where reliable estimates could be calculated from 1 to 3 percent of the analyses (the Czech Republic, Hong Kong SAR, and the United States), and the other two were not found to differ from all the other countries on any feature examined in this study (Australia and Switzerland).

Among the 17 analyses on which Japan differed from all of the other countries, most were related to the mathematical problems that were worked on during the lesson and to instructional practices. For example, Japanese eighth-grade lessons were found to be higher than all the other countries where reliable estimates could be calculated on the percentage of lesson time spent introducing new content (60 percent, figure 3.8), the percentage of high complexity problems (39 percent, figure 4.1), mathematically related problems (42 percent, figure 4.6), problems that were repetitions (40 percent, figure 4.6), problems that were summarized (27 percent, table 5.4), and the average time spent on a problem (15 minutes, figure 3.5). In some cases, Japanese lessons were found to be higher than lessons in the other countries on a feature, and in other cases, lower.

Dutch eighth-grade mathematics lessons, on the other hand, differed from all the other countries on 8 analyses, mostly related to the structure of the lesson and particular instructional practices. For example, Dutch lessons were higher than all the other countries where reliable estimates could be calculated on the percentage of lessons that contained goal statements (21 percent, figure 3.12) and in which a calculator was used (91 percent, figure 5.18), the percentage of problems per lesson that were set up using only mathematical language or symbols (40 percent, figure

---

[1] This does not take into account differences between participating countries based on teachers' reports of years teaching, academic preparation, lesson goals, and the like, as reported in chapter 2.

5.1), and the estimated average time per lesson spent on future homework problems (10 minutes, table 3.8). Like Japan, in some cases, Dutch eighth-grade mathematics lessons were found to be higher than lessons in the other countries on a feature, and in other cases, lower.

These results show that some features of eighth-grade mathematics teaching in several countries, particularly in Japan and the Netherlands, were not shared by the other countries. Although these differences are limited to a few features relative to the number analyzed for this report, the criterion for being "unique" is set quite high by requiring a significant difference from all the other countries. The number of features on which differences were found rises if the question becomes whether an individual country differed from the majority of the other countries. Looking at the issue of relative differences another way, it was found that on 5 percent of the analyses conducted no differences were found among any of the participating countries. When a close-up lens is used, differences in teaching practices become evident, especially on features that describe the mathematics problems presented and the ways in which they were worked on during the lesson.

Looking Through Both Lenses, and Asking More Questions

Is there a way to reconcile these findings that suggest there are both broad similarities in eighth-grade mathematics lessons across the countries as well as particular differences across the countries in how mathematics problems are presented and solved? One approach is to notice that similar ingredients or building blocks can be combined in different ways to create different kinds of lessons. Teaching across countries simultaneously can look both similar and different, (Anderson-Levitt 2002).

It is often the case that when mathematics teaching and learning are compared across countries, it is the differences that receive particular attention (Leung 1995; Manaster 1998; Schmidt et al. 1999; Stigler and Hiebert 1999). In this study, as well, the differences among the countries are intriguing because, in part, they are the kinds of differences that might relate to different learning experiences for students (Clarke 2003; National Research Council 2001). In particular, differences in the kinds of mathematics problems presented and how lessons are constructed to engage students in working on the problems can yield differences in the kinds of learning opportunities available to students (National Research Council 2001; Stein and Lane 1996).

How can the differences in teaching among the countries be explored further? The previous chapters, and the examples reviewed above, reported differences between pairs of countries feature by feature. Beyond comparing countries on individual features, there are at least two approaches that can provide additional insights: one approach is to look inside each country, at constellations of features, that describe the way in which lessons were constructed; and a second approach is to re-examine individual features within the context of each country's "system" of teaching. Both of these approaches are explored in the next section.

What Are the Relationships Among Features of Mathematics Teaching in Each Country?

Lesson Signatures in Each Country

If there are features that characterize teaching in a particular country, there should be enough similarities across lessons within the country to reveal a particular pattern to the lessons in each country. If this were the case, then overlaying the features of all of the lessons within a country would reveal a pattern or, as labeled here, a "lesson signature."

The analyses presented to this point in the report focused on the presence of particular lesson features. In contrast, lesson signatures take into account when features occurred in the course of a lesson and consider whether and how basic lesson features co-occurred. The lesson signatures shown below were created by considering three dimensions that provide a dynamic structure to lessons: the purpose of the activities, the type of classroom interaction, and the nature of the content activity. These three dimensions were comprised of selected features analyzed for this study, and generally follow the organization of the three main chapters in this report (chapters 3 through 5). To create a lesson signature, each eighth-grade mathematics lesson was exhaustively subdivided along each of the three dimensions by marking the beginning and ending times for any shifts in the features. In this way, the dimensions could be linked by time through the lesson. This allowed an investigation into the ways in which the purpose segments, classroom interaction phases, and content activities appeared, co-occurred, and changed as the lessons proceeded.

The lesson signatures presented on the following pages were constructed by asking what was happening along the three dimensions during each minute of every eighth-grade mathematics lesson.[2] Each variable or feature within a dimension is listed separately and is accompanied by its own histogram which represents the frequency of occurrence across all the lessons in that country, expressed as a percentage of the eighth-grade mathematics lessons. The histogram increases in height by one pixel[3] for every 5 percent of lessons marked positively for a feature at any given moment during the lesson time, and disappears when fewer than 5 percent of lessons were marked positively (due to technological limitations).

Along the horizontal axis of each lesson signature is a time scale that represents the percentage of time that has elapsed in a given lesson, from the beginning to the end of a lesson. The percentage of lesson time was used to standardize the passing of time across lessons which can vary widely in length, from as little as 28 minutes to as much as 119 minutes (see table 3.1). Representing the passing of time in this way provides a sense of the point in a lesson that an activity or event occurred relative to the point in another lesson that the same activity or event occurred. For example, if lesson A was twice as long as lesson B, and the first mathematics problem in lesson A was presented 6 minutes into the lesson and the first mathematics problem in lesson B was presented 3 minutes into the lesson, the lesson signature would show that the first mathematics problem in both lessons occurred at the same relative time.

---

[2]The analysis used to develop the lesson signatures divided each lesson into 250 segments, each representing 0.4 percent of the total lesson length. For example, the analysis accounts for how a 50-minute lesson was coded approximately every 12 seconds.
[3]Pixel is short for "picture element." A pixel is the smallest unit of visual information that can be used to build an image. In the case of the printed page, pixels are the little dots or squares that can be seen when a graphics image is enlarged or viewed up close.

To assist the reader in gauging the passing of time in the lessons, each lesson signature has vertical lines marking the beginning (zero), middle (50 percent), and end (100 percent) of the lesson time, moving from left to right. The 20, 40, 60, and 80 percent marks are indicated as well. By following the histogram of a particular feature from the zero to the 100 percent of time markings, one can get a rough idea of the percentage of lessons that included the feature at various moments throughout the lesson. For example, a lesson signature may show that 100 percent of lessons begin with review, but by the midpoint of a lesson, the percentage of lessons that are focused on review has decreased. As an additional aid to the reader, tables that list the percentage of lessons that included each feature from the zero to 100 percent time marks (in increments of 20) are included in appendix F.

Comparing the histograms of features within or across dimensions provides a sense of how those features were implemented as lesson time elapsed. Patterns may or may not be easily identified. Where patterns are readily apparent, this suggests that many lessons contain the same sequence of features. Where patterns are not readily apparent, this suggests variability within a country, either in terms of the presence of particular features or in terms of their sequencing. Furthermore, if the histograms of particular features are all relatively high at the same time in the lesson, this suggests that these features are likely to co-occur. However, in any single lesson observed in a country, this may or may not be the case. Thus, the histograms provide a general sense of what occurs as lesson time passes rather than explicitly documenting how each lesson moved from one feature to the next.

As noted above, a set of features within each of three dimensions are displayed along the left side of the lesson signatures (i.e., purpose, classroom interaction, and content activity). Within each dimension, the features that are used to represent each dimension are mutually exclusive (that is, a lesson was coded as exhibiting only one of the features at any point in time). However, in the interest of space, some low frequency features in two of the dimensions are not shown. For classroom interaction, the features not shown are "optional, teacher presents information" and "mixed private and public interaction" (these two features, combined, accounted for no more than 2 percent of lesson time in each country; see table 3.6). For content activity, the feature not shown is "non-mathematical work" (accounting for no more than 2 percent of lesson time in each country; see figure 3.2).

All of the features presented in the lesson signatures are defined and described in detail in chapters 3, 4, and 5. The lessons signatures show additional detail about independent and concurrent problems. As stated earlier, independent problems were presented as single problems and worked on for a clearly recognizable period of time. Concurrent problems were presented as a set of problems that were worked on privately for a time. To provide the reader with a sense of the utilization of independent problems in the lessons, independent problems are grouped into 4 categories: the first independent problem worked on in the lesson, the second through fifth independent problems worked on in the lesson, the sixth through tenth independent problems worked on in the lesson, and the eleventh and higher independent problems worked on in the lesson. For concurrent problems, it was possible to distinguish between times when they were worked on through whole-class, public discussion (concurrent problems, classwork) and times when they were worked on through individual or small-group work (concurrent problems, seatwork). These two features are displayed in the lesson signatures as well.

Each lesson signature provides a view, at a glance, of how the lessons from a country were coded for each of the three dimensions (i.e., purpose, classroom interaction, and content activities).

The lesson signature for each country will be discussed in turn, and will also be supplemented by findings presented in prior chapters. In this way, the lesson signatures become a vehicle for pulling together the many pieces of information contained in this report. Because each signature displays 15 histograms, it is often difficult to assess the exact frequency of a given code at a particular moment in the lesson. As stated earlier, percentages for each feature, organized by country, are included in tables in appendix F.

The lesson signature for Australia

Purpose

As stated earlier, 89 percent of eighth-grade Australian mathematics lessons included some portion of time during the class period devoted to review (table 3.4), representing an average of 36 percent of time per lesson (figure 3.8). Moreover, 28 percent of mathematics lessons were found to spend the entire lesson time in review of previously learned content, among the highest percentages of all the countries (figure 3.9). As visible on the lesson signature (figure 6.1), 87 percent of the Australian eighth-grade mathematics lessons began with a review of previously learned content. A majority of Australian lessons focused on review through the first 20 percent of lesson time, with a decreasing percentage of lessons going over previously learned content through the remainder of the lesson (figure 6.1 and table F.1, appendix F). Starting at about one-third of the way into the lesson and continuing to the end, a majority of Australian lessons engaged students with new content, representing an average of 56 percent of time per lesson (figure 3.8), with the practicing of new content becoming an increasing focus in the latter parts of the lesson (figure 6.1 and table F.1, appendix F).

Classroom interaction

In terms of the interaction format in which eighth-grade students and teachers worked on mathematics, Australian lessons were found to show no detectable difference in the percentage of lesson time devoted to whole-class, public interaction versus private, individual or small-group interaction (52 and 48 percent, on average, respectively, table 3.6). The majority of Australian lessons were conducted through whole-class, public interaction during roughly the first third of lesson time, and again at the very end of the lesson (figure 6.1 and table F.1, appendix F). In between those two periods of time in the lesson, eighth-grade Australian students were found to be engaged in private work in a majority of lessons, usually with students working individually on problems that asked students to repeat procedures that had been demonstrated earlier in the lesson (75 percent of private interaction time per lesson was spent working individually, on average, figure 3.10; 65 percent of student private work time is spent repeating procedures that had been demonstrated earlier in the lesson, figure 5.13). In the hypothesized Australian country model, experts posited that there would be a "practice/application" period in the typical eighth-grade Australian mathematics lesson during which students would often be assigned a task to complete privately while the teacher moved about the class assisting students (figure E.2, appendix E). As the lesson signature shows, at a time in the lesson when the largest percentage of Australian lessons were focused on the practice of new content (roughly during most of the last third of the lesson), a majority of Australian lessons were found to have students working individually or in small groups (private interaction; figure 6.1 and table F.1, appendix F).

Content activities

During the first half of the Australian mathematics lessons, there does not appear to be any consistent pattern in the types of problems that are presented to students (figure 6.1 and table F.1, appendix F). That is, teachers conveyed previously learned or new content to students by working on independent problems, sets of problems, either as a whole class or as seatwork (concurrent problems), or on mathematics outside the context of a problem (e.g., presenting definitions or concepts, pointing out relationships among ideas, or providing an overview of the lesson). During most of the last half of the Australian eighth-grade mathematics lessons, however, a majority of lessons were found to employ sets of problems (concurrent problems) as a way to focus on new content.

The delivery of content in Australian lessons is also revealed in analyses presented earlier in the report but not readily evident in the lesson signature. For example, when taking into consideration all of the problems presented in the eighth-grade Australian mathematics lessons, except for answered-only problems, 61 percent of problems per lesson were found to be posed by the teacher with the apparent intent of using procedures—problems that are typically solved by applying a procedure or set of procedures. This is a higher percentage than the percentage posed by the teacher with the apparent intent of either making connections between ideas, facts, or procedures, or to elicit a mathematical convention or concept (stating concepts; 24 and 15 percent, respectively, figure 5.8). Furthermore, when the problems introduced in the lesson were examined a second time for processes made public while working through the problems, 77 percent of the problems per lesson in Australia were found to have been solved by focusing on the procedures necessary to solve the problem or by simply giving results only without discussion of how the answer was obtained (figure 5.9). Moreover, when the 15 percent of problems per lesson that were posed to make mathematical connections were followed through to see whether the connections were stated or discussed publicly, 8 percent per lesson were solved by explicitly and publicly making the connections (figure 5.12). Finally, when experts examined the problems worked on or assigned during each lesson for the level of procedural complexity—based on the number of steps it takes to solve a problem using common solution methods—77 percent of the problems per eighth-grade mathematics lesson in Australia were found to be of low complexity, among the highest percentages in all the countries (figure 4.1).

These observations and findings suggest that, on average, eighth-grade Australian mathematics lessons were conducted through a combination of whole-class, public discussion and private, individual student work, with an increasing focus on students working individually on sets of problems that were solved by using similar procedures as new content was introduced into the lesson and practiced.

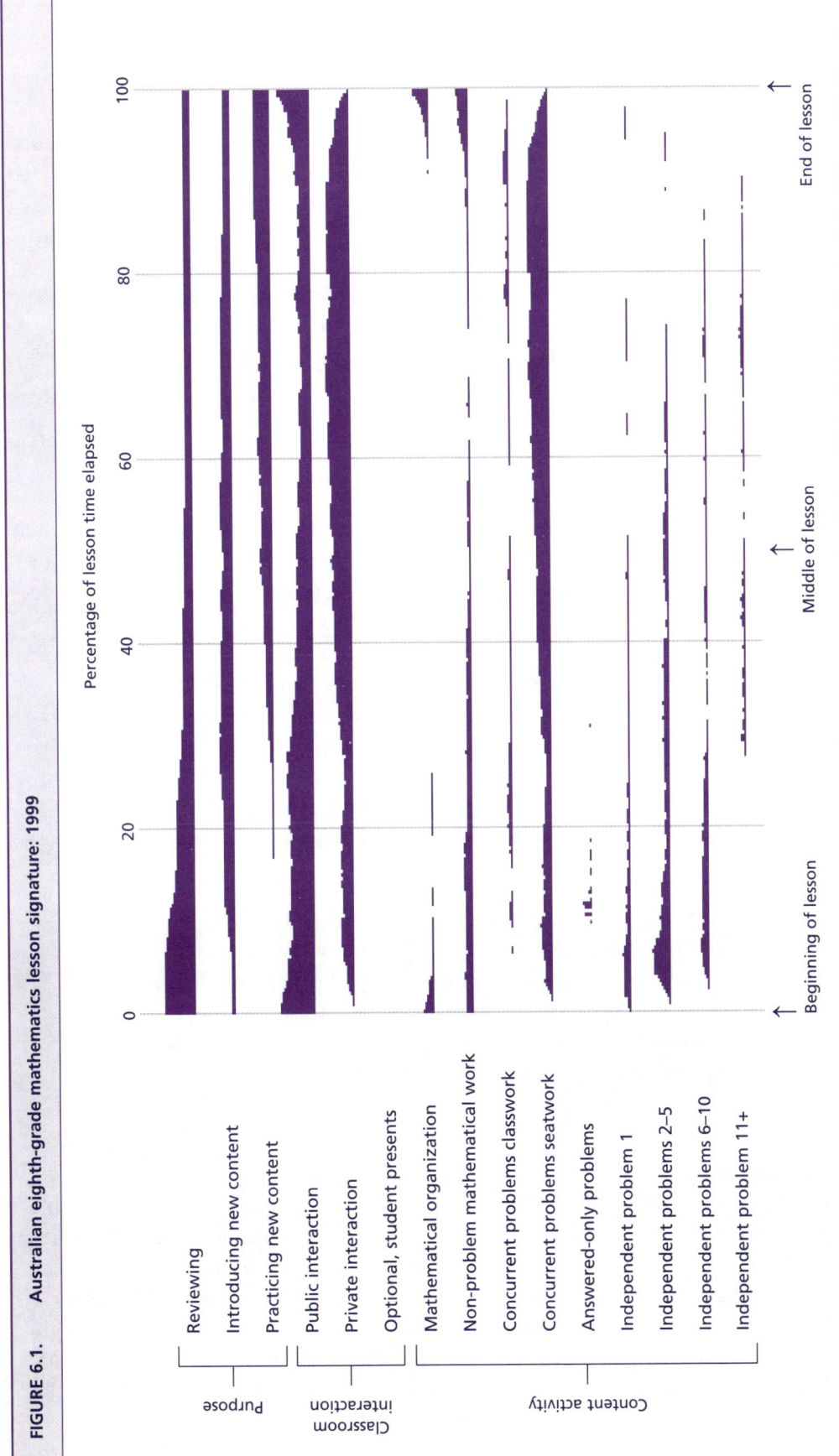

Chapter 6 | 127
Similarities and Differences in Eighth-Grade Mathematics Teaching Across Seven Countries

FIGURE 6.1. Australian eighth-grade mathematics lesson signature: 1999

NOTE: The graph represents both the frequency of occurrence of a feature and the elapsing of time throughout a lesson. For each feature listed along the left side of the graph, the histogram (or bar) represents the percentage of eighth-grade mathematics lessons that exhibited the feature—the thicker the histogram, the larger the percentage of lessons that exhibited the feature. From left to right, the percentage of elapsed time in a lesson is marked along the bottom of the graph. The histogram increases by one pixel (or printable dot) for every 5 percent of lessons marked for a feature at any given moment during the lesson time, and disappears when fewer than 5 percent of lessons were marked (due to technological limitations). By following each histogram from left to right, one can get an idea of the percentage of lessons that included the feature as lesson time elapsed. A listing of the percentage of lessons that included each feature by the elapsing of time is included in appendix F. To create each histogram, each lesson was divided into 250 segments, each representing 0.4 percent of lesson time. The codes applied to each lesson at the start of each segment were tabulated, using weighted data, and reported as the percentage of lessons exhibiting each feature at particular moments in time.
SOURCE: U.S. Department of Education, National Center for Education Statistics, Third International Mathematics and Science Study (TIMSS), Video Study, 1999.

The lesson signature for the Czech Republic

Purpose

One of the distinguishing characteristics of eighth-grade mathematics teaching in the Czech Republic was the relatively prominent role of review, as illustrated in the lesson signature (figure 6.2). Support for this statement is based on the observation that a majority of Czech mathematics lessons focused on the review of previously learned content throughout the first half of the lesson, particularly during roughly the first third of the lesson (table F.2, appendix F). Findings presented in prior chapters also indicate that 100 percent of eighth-grade Czech mathematics lessons contain at least some portion of the lesson devoted to review (table 3.4), consuming an average of 58 percent of time per lesson, surpassing the average in all the countries except the United States (figure 3.8). Moreover, one-quarter of Czech lessons were found to spend the entire lesson time in review (figure 3.9).

The hypothesized Czech country model (figure E.3, appendix E), as compiled by country experts, suggests that review is not intended as a time to go over homework. Indeed, an earlier analysis shows that, on average, less than one minute per lesson was spent reviewing homework (table 3.9). Rather, according to the hypothesized country model, review in Czech lessons includes such goals as re-instruction and securing old knowledge, and serves as an opportunity for teachers to evaluate students. As noted in chapter 3, this latter point was observed in the Czech mathematics lessons during the interaction pattern of students presenting information that was optional for other students, an interaction pattern that distinguished the eighth-grade Czech mathematics lessons from lessons in all the other countries where reliable estimates could be calculated (table 3.6). As shown in the hypothesized Czech country model and as observed in the videotaped lessons, oral examinations could be given at the beginning of the lesson. In these cases, one or two students were called to the board to solve a mathematical problem presented by the teacher, for which the students were then publicly graded on their performance. As also observed in the videotapes, the students could be asked to describe in detail the steps to solve the problem.

Though review appears to have played a prominent role in eighth-grade Czech mathematics lessons, as can be observed in the lesson signature, the focus in a majority of the lessons turned to the introduction and practice of new content after the midpoint of the lesson (figure 6.2 and table F.2, appendix F). Nonetheless, an examination of the way in which lesson time is divided among the three purposes defined for this study revealed that eighth-grade Czech lessons devoted a greater proportion of lesson time to review than to introducing and practicing new content combined (58 percent compared to 42 percent, on average, figure 3.8).

Classroom interaction

The way in which students and teachers interacted during the lesson in eighth-grade Czech mathematics lessons was characterized by the predominance of whole-class, public interaction (figure 6.2 and table F.2, appendix F). Indeed, 61 percent of lesson time, on average, was spent in public interaction (table 3.6). Conversely, the lesson signature shows the relative infrequency of students working individually or in small groups (private interaction). During the 21 percent of average lesson time that students spent in private interaction (table 3.6), it was overwhelmingly organized for students to work individually rather than in small groups or pairs (92 percent of private interaction time per lesson, on average, was spent working individually, figure 3.10).

As already noted above, the eighth-grade Czech mathematics lessons exhibited a third interaction pattern that was unique in comparison to the other countries: the public presentation of information by a student in response to the teacher's request, during which time the other students could interact with that student and teacher, or work on an assignment at their desk (optional, student presents information). On average, 18 percent of lesson time in Czech lessons was devoted to this form of interaction (table 3.6), though as is shown in the lesson signature, this form of classroom interaction was not characteristic to any particular point in a lesson (figure 6.2 and table F.2, appendix F).

Content activities

As seen in the Czech lesson signature (figure 6.2), a noticeable percentage of eighth-grade Czech mathematics lessons began and ended with non-problem segments—that is, segments in which mathematical information was presented but problems were not worked on (see also table F.2, appendix F). Given that 91 percent of Czech mathematics lessons contained a goal statement by the teacher that made clear what would be covered during the lesson (figure 3.12), this may account, in part, for the 45 percent of lessons that began with non-problem segments (table F.2, appendix F).

When considering the ways in which mathematics problems were worked on during the lesson, the lesson signature reveals that a majority of lessons utilized independent problems to engage students in mathematics, throughout most of the lesson time (except at the very beginning and end of the lessons; table F.2, appendix F). Though the use of sets of problems (concurrent problems) is not uncommon in eighth-grade Czech mathematics lessons, analyses revealed that, on average, Czech lessons included 13 discrete, independent problems, among the highest number in all the countries (table 3.3) and constituting an average of 52 percent of the total lesson time (figure 3.4). Moreover, when problems were introduced into the lesson, 81 percent of problems per lesson were set up using mathematical language or symbols only (figure 5.1) and 77 percent were posed by the teacher with the apparent intent of using procedures—problems that are typically solved by applying a procedure or set of procedures (figure 5.8). This latter finding was found to be higher than the percentage of problems per lesson that were posed to make connections between ideas, facts, or procedures, or problems that were posed to elicit a mathematical convention or concept (stating concepts; 16 and 7 percent, on average, respectively, figure 5.8). When the problems introduced in the lesson were examined a second time for the processes made public while working through the problems, 71 percent of problems per lesson were found to have been solved by simply giving results only without discussion of how the answer was obtained or by focusing on the procedures necessary to solve the problem (figure 5.9). Moreover, when the 16 percent of problems per lesson that were posed to make mathematical connections were followed through to see whether the connections were stated or discussed publicly, 52 percent per lesson were solved by explicitly and publicly making the connections (figure 5.12). Finally, when experts examined the problems worked on or assigned during each lesson for the level of procedural complexity—based on the number of steps it takes to solve a problem using common solution methods—64 percent of the problems per eighth-grade mathematics lesson in the Czech Republic were found to be of low complexity, 25 percent of moderate complexity, and 11 percent of high complexity (figure 4.1).

Looking back across these findings suggests that, on average, eighth-grade Czech mathematics lessons emphasized review to, in part, check students' knowledge and conveyed new content by having students work on a relatively high number of independent problems that required a focus on using procedures, all of which was conducted largely through whole-class, public interactions.

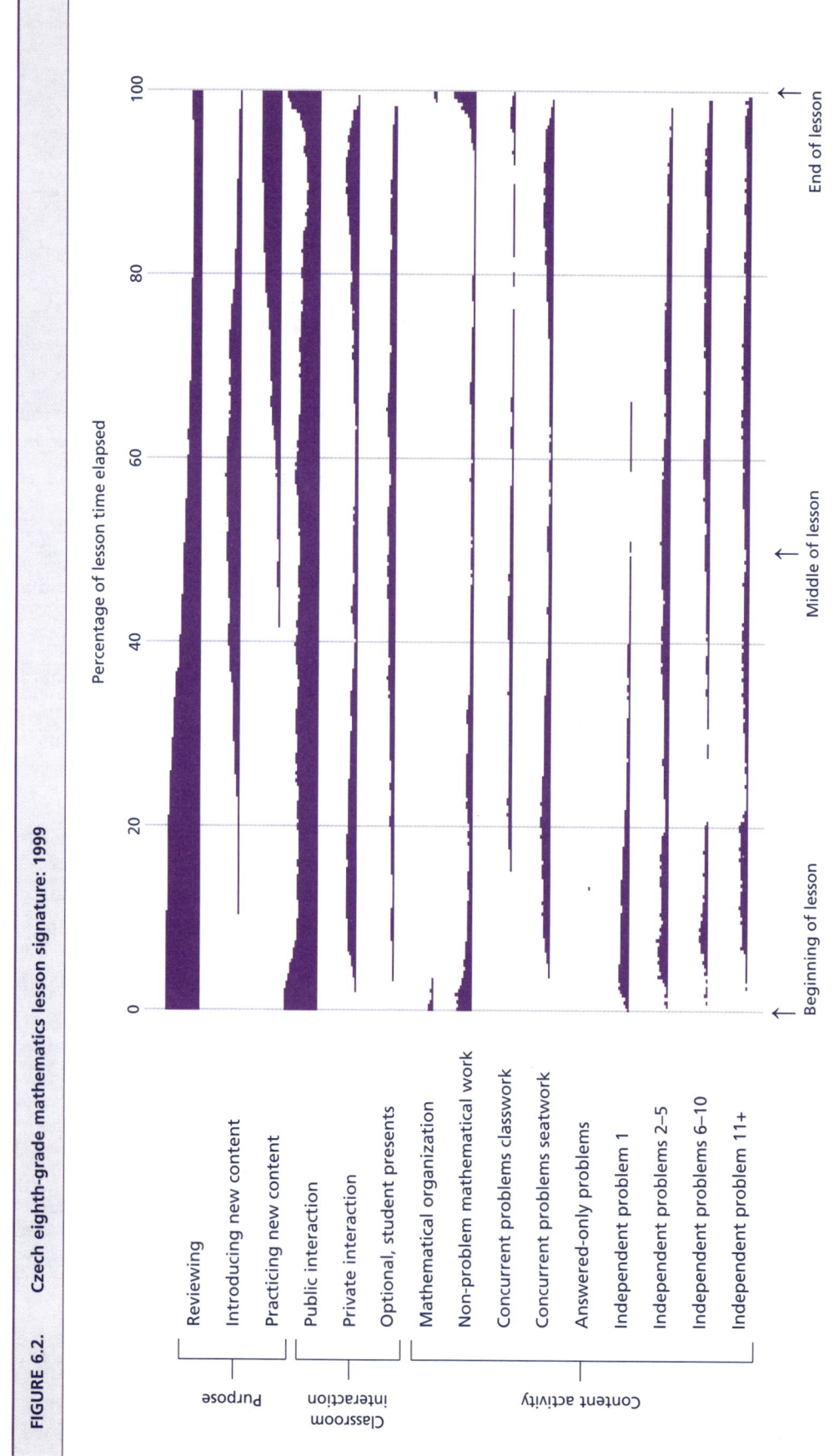

130 | Teaching Mathematics in Seven Countries
Results From the TIMSS 1999 Video Study

FIGURE 6.2. Czech eighth-grade mathematics lesson signature: 1999

NOTE: The graph represents both the frequency of occurrence of a feature and the elapsing of time throughout a lesson. For each feature listed along the left side of the graph, the histogram (or bar) represents the percentage of eighth-grade mathematics lessons that exhibited the feature—the thicker the histogram, the larger the percentage of lessons that exhibited the feature. From left to right, the percentage of elapsed time in a lesson is marked along the bottom of the graph. The histogram increases by one pixel (or printable dot) for every 5 percent of lessons marked for a feature at any given moment during the lesson time, and disappears when fewer than 5 percent of lessons were marked (due to technological limitations). By following each histogram from left to right, one can get an idea of the percentage of lessons that included the feature as lesson time elapsed. A listing of the percentage of lessons that included each feature by the elapsing of time is included in appendix F. To create each histogram, each lesson was divided into 250 segments, each representing 0.4 percent of lesson time. The codes applied to each lesson at the start of each segment were tabulated, using weighted data, and reported as the percentage of lessons exhibiting each feature at particular moments in time.

SOURCE: U.S. Department of Education, National Center for Education Statistics, Third International Mathematics and Science Study (TIMSS), Video Study, 1999.

The lesson signature for Hong Kong SAR

Purpose

In Hong Kong SAR, 24 percent of eighth-grade mathematics lesson time, on average, was devoted to review, among the lowest percentages of all the countries (figure 3.8). Nonetheless, analysis of the data shows that 82 percent of lessons contained a review segment (table 3.4). The lesson signature (figure 6.3) shows that when review was included as part of the lesson, it tended to occur in the early moments of the lesson (see also table F.3, appendix F). Indeed, although about three-quarters (77 percent) of the lessons initially began with a review of previously learned content, one-third (33 percent) focused on review at the 20 percent mark of elapsed time (table F.3, appendix F). In the hypothesized Hong Kong SAR country model developed by country experts, review time was mentioned as preparation for the new content to be presented later in the lesson (figure E.4, appendix E). A specific routine described by Hong Kong SAR experts was the process of going over content presented in past lessons that was relevant to the day's new procedures or concepts.

Eighth-grade mathematics lessons in Hong Kong SAR were among those lessons that, on average, spent the largest percentage of time on introducing and practicing new content (76 percent, figure 3.8), which is visible in the lesson signature (figure 6.3). Indeed, 92 percent of eighth-grade mathematics lessons in Hong Kong SAR were found to contain a segment in which new content was introduced in the lesson, among the highest percentages of all the countries (table 3.4). Conversely, the percentage of lessons that were devoted entirely to review in Hong Kong SAR was among the lowest of the countries (8 percent, figure 3.9). All of this would seem to suggest that, although a large percentage of eighth-grade Hong Kong SAR mathematics lessons included review, it played a relatively lesser role in the lessons in terms of the proportion of lesson time devoted to the activity.

Finally, although about three-quarters of lesson time in Hong Kong SAR mathematics classes was devoted to new content, beginning around the midpoint of the lesson, the practice of new content introduced in the lesson became an increasing focus of activity (figure 6.3 and table F.3, appendix F). Country experts highlighted the role of practicing new content in the hypothesized Hong Kong SAR country model (figure E.4, appendix E). The term used by experts to label the practicing phase was "consolidation." According to the experts, consolidation is accomplished through private work, the assessment of student work (i.e., some students working at the board), or homework. Country experts indicated that through the practice phase, teachers aim to increase students' confidence that new problems can be completed. According to these experts, the practice phase ensures that homework problems can be worked successfully and that the skills can be applied accurately on future examinations. It should be recalled that the average duration of a mathematics lesson in Hong Kong SAR was the shortest among all the countries (table 3.1), suggesting that activities are likely to have been compressed and focused.

Classroom interaction

As supported by the nearly solid band that represents public interaction in figure 6.3, an average of three-quarters of mathematics lesson time in Hong Kong SAR was spent in public interaction, a greater proportion of lesson time than in all the other countries except the United States (table 3.6). During this time, the teacher talked much more than the students, in a ratio of teacher to student words of 16:1 (figure 5.15). When students contributed verbally, it was often limited to brief utterances (figure 5.17). Country experts suggested that one possible interpretation of the high percentage of teacher talk, along with the emphasis on introducing and practicing new content, is a press for efficiency. The experts suggested that teachers and students share an expectation that the relatively short lesson time (an average of 41 minutes and a median of 36 minutes, table 3.1) is designed to cover new content for which students will be accountable in the future.

That three-quarters of lesson time, on average, was spent in whole-class, public discussion means that an average of one-quarter of lesson time was spent in some other form of classroom interaction. As reported earlier, an average of 20 percent of lesson time in eighth-grade Hong Kong SAR mathematics lessons was devoted to students working largely independently (table 3.6 and figure 3.10). Unlike in some of the other countries in the study in which a majority of mathematics lessons turned toward individual or small-group work during some continuous portion of the lesson (i.e., Australia, the Netherlands, and Switzerland), Hong Kong SAR was among the countries that engaged students largely through whole-class, public interaction (i.e., the Czech Republic, Japan, and the United States; see tables 3.6 and F.1 through F.7, appendix F).

Content activities

An examination of the lesson signature shows that in addition to a number of eighth-grade mathematics lessons in Hong Kong SAR starting out with a review of previously learned content (77 percent of lessons start out this way), a number also initially focused on content activities such as non-problem-based mathematical work (43 percent): presenting definitions, pointing out relationships among ideas, or providing an overview or summary of the major points in a lesson. Moreover, although 20 percent of Hong Kong SAR lessons began by working on an independent mathematics problem, by roughly one fifth of the way through the lesson 67 percent focused on independent problems (figure 6.3 and table F.3, appendix F). This point in the lesson coincided with a shift from review of previously learned content to the introduction of new content in a majority of the lessons (58 percent of lessons were found to focus on the introduction of new content at around the same time point; figure 6.3 and table F.3, appendix F). As students and teachers moved through the second half of the lesson, Hong Kong SAR mathematics lessons also began to focus on the practice of new content through a mix of independent problems and sets of problems (concurrent problems) assigned to students as whole-class or seatwork (figure 6.3 and table F.3, appendix F).

The content of the mathematics introduced into eighth-grade Hong Kong SAR lessons was also shaped, in part, by the format of the mathematics problems and the ways in which the students worked on the problems. For example, 83 percent of problems per lesson in Hong Kong SAR were conveyed through the use of mathematical language or symbols only (figure 5.1), among the highest percentages of all the countries. When the problems introduced into eighth-grade mathematics lesson in Hong Kong SAR were examined by experts for the level of procedural complexity—based on the number of steps it takes to solve a problem using common solution methods—63 percent of problems per lesson were found to be of low procedural complexity, 29 percent of moderate complexity, and 8 percent of high complexity (figure 4.1). Moreover, analyses revealed that 84 percent of problems per lesson were posed by the teacher with an apparent intent of using procedures—problems that are typically solved by applying a procedure or set of procedures—among the highest percentages of all the countries (figure 5.8). When all the mathematics problems were examined a second time to understand the processes publicly discussed by teachers and students as they worked through the problems, 48 percent of problems per eighth-grade mathematics lesson in Hong Kong SAR were found to have focused publicly on the procedures needed to solve the problem and 15 percent were solved by giving results only without discussion of how the answer was obtained (figure 5.9). Finally, analyses revealed that 13 percent of problems in Hong Kong SAR mathematics lessons were posed by the teacher with an apparent intent of making connections—problems that are typically solved by constructing relationships between mathematical ideas (figure 5.8). When these problems were examined a second time to understand the processes publicly discussed by teachers and students as they worked through them, on average, 46 percent were found to have focused publicly on making connections (figure 5.12).

These findings suggest that during the relatively short period of time spent on mathematics in Hong Kong SAR, eighth-grade lessons, on average, emphasized the introduction and practice of new content through whole-class, public discussion and working on problems that focused on using procedures.

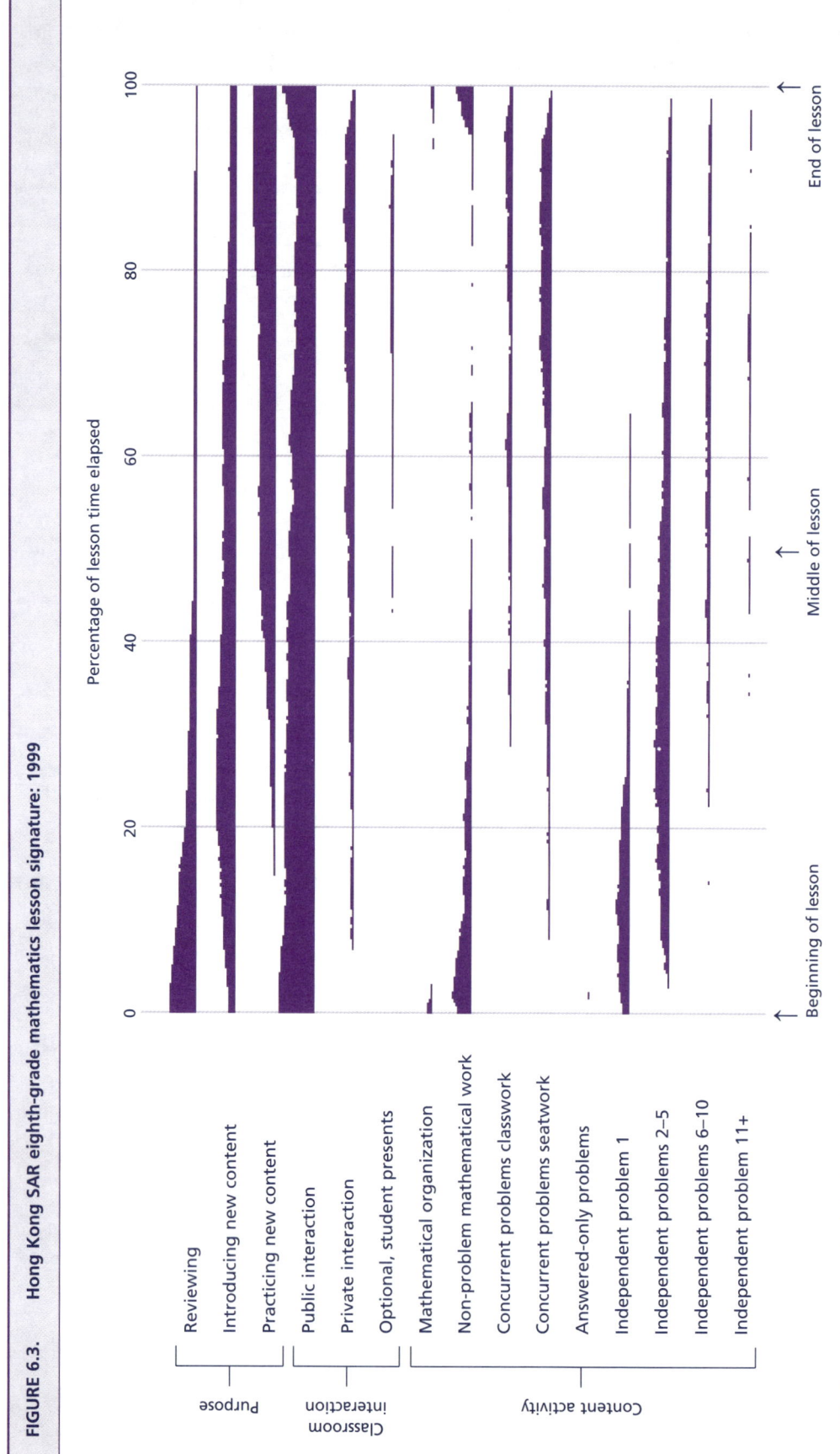

134 | Teaching Mathematics in Seven Countries
Results From the TIMSS 1999 Video Study

FIGURE 6.3. Hong Kong SAR eighth-grade mathematics lesson signature: 1999

NOTE: The graph represents both the frequency of occurrence of a feature and the elapsing of time throughout a lesson. For each feature listed along the left side of the graph, the histogram (or bar) represents the percentage of eighth-grade mathematics lessons that exhibited the feature—the thicker the histogram, the larger the percentage of lessons that exhibited the feature. From left to right, the percentage of elapsed time in a lesson is marked along the bottom of the graph. The histogram increases by one pixel (or printable dot) for every 5 percent of lessons marked for a feature at any given moment during the lesson time, and disappears when fewer than 5 percent of lessons were marked (due to technological limitations). By following each histogram from left to right, one can get an idea of the percentage of lessons that included the feature as lesson time elapsed. A listing of the percentage of lessons that included each feature by the elapsing of time is included in appendix F. To create each histogram, each lesson was divided into 250 segments, each representing 0.4 percent of lesson time. The codes applied to each lesson at the start of each segment were tabulated, using weighted data, and reported as the percentage of lessons exhibiting each feature at particular moments in time.
SOURCE: U.S. Department of Education, National Center for Education Statistics, Third International Mathematics and Science Study (TIMSS), Video Study, 1999.

The lesson signature for Japan

Purpose

As noted earlier, some features of eighth-grade mathematics teaching in Japan were different from all the other participating countries. The lesson signature for Japan displays another way of looking at some of these features. Although 73 percent of Japanese eighth-grade mathematics lessons began with a review of previously learned content (figure 6.4 and table F.4, appendix F), the review of previously learned content constituted an average of 24 percent of lesson time (figure 3.8), among the smallest percentages in all of the countries. Indeed, as the lesson signature shows, by the time that 20 percent of the lesson time had passed, a majority of Japanese mathematics lessons were focused on the introduction of new content (figure 6.4 and table F.4, appendix F), which continued throughout the remaining lesson time. This relative emphasis on the introduction of new content in Japanese lessons is supported by two previously reported analyses: 95 percent of lessons included some portion of lesson time to introduce new content (table 3.4), averaging 60 percent of time per lesson (figure 3.8), the largest proportion of time among all the countries. Although the introduction of new content seemed to be a primary focus in the Japanese lessons, starting at the midpoint of the lesson, an increasing percentage of lessons moved toward the practice of the new content, constituting an average of 16 percent of time per lesson (figure 3.8).

Classroom interaction

Like lessons in several of the other countries that participated in this study, eighth-grade Japanese mathematics lessons were conducted largely through whole-class, public discussions (figure 6.4 and table F.4, appendix F). A majority of Japanese mathematics lessons were carried out in this way throughout most of the lesson time, averaging 63 percent of time per lesson to this form of classroom interaction (table 3.6). During the roughly one-third of lesson time that was not characterized by whole-class instruction, students worked largely on their own (table 3.6 and figure 3.10). Throughout the course of a lesson, eighth-grade Japanese mathematics lessons were found to shift back and forth between whole-class instruction and individual student work an average of 8 times per lesson, more often than any of the other countries except the Czech Republic (table 3.7). Thus, Japanese mathematics teachers varied the type of interaction taking place in the classroom more often than in any of the other countries except the Czech Republic.

Content activities

Observation of the eighth-grade Japanese mathematics lessons suggests that the teachers conduct the lessons by focusing on a relatively few number of independent problems related to a single topic for extended periods of time. This can be seen in the Japanese lesson signature in the focus of a majority of lessons on one to five problems over the course of a lesson (figure 6.4 and table F.4, appendix F). On average, eighth-grade Japanese mathematics lessons introduced three independent problems per lesson, fewer than all of the other countries except Australia (table 3.3), taking up an average of 64 percent of lesson time (figure 3.4). Moreover, devoting a large percentage of time to work on a relatively few problems also meant that the average time spent per problem was uniquely long in Japan—15 minutes on average (figure 3.5). Ninety-four percent of eighth-grade mathematics lessons in Japan included problems that related to a single topic, as opposed to more than one topic (figure 4.8). When problems were introduced into the lesson, 89 percent of the problems per lesson were set up using only mathematical language or symbols whereas 9 percent of the problems contained references to real-life contexts, among the smallest percentages of the countries (figure 5.1). Of all the problems introduced into the eighth-grade Japanese mathematics lessons, 41 percent per lesson were posed with the apparent intent of using procedures, among the smallest percentages of all the countries (figure 5.8). The majority of problems per lesson were posed with the apparent intent of making connections between ideas, facts, or procedures (54 percent, figure 5.8). When the problems introduced in the lessons were examined a second time for the processes that were made public while working through the problems, 27 percent of problems per lesson were solved by explicitly using procedures and 37 percent were solved by making explicit the relevant mathematical connections (figure 5.9). Moreover, when the 54 percent of problems per lesson that were posed to make mathematical connections were followed through to see whether the connections were stated or discussed publicly, almost half of the problems per lesson (48 percent) were solved by explicitly making the connections (figure 5.12).

Other indications of how eighth-grade mathematics lessons in Japan are conducted show that 17 percent of problems per lesson, on average, included a public discussion and presentation of alternative solution methods, among the highest percentages of all the countries (table 5.1). Fifteen percent of problems per mathematics lesson in Japan were accompanied with a clear indication by the teacher that students could choose their own method for solving the problem (table 5.2). About one-quarter (27 percent) of mathematics problems per lesson were summarized by the teacher to clarify the mathematical point illustrated by the problem (table 5.4), more than in any of the other countries. Finally, based on an examination by experts into the level of procedural complexity of the problems introduced into the Japanese lesson—by looking at the number of steps it takes to solve a problem using common solution methods—17 percent of problems per lesson were found to be of low procedural complexity, 45 percent of moderate complexity, and 39 percent of high complexity (figure 4.1).[4]

Looking back across these observations and findings suggests that on average eighth-grade Japanese mathematics lessons were conducted largely through whole-class, public discussion during which the emphasis was on the introduction of new content, conveyed to students by focusing on a few number of independent problems related to a single topic over a relatively extended period of time, with the goal of making connections among mathematical facts, ideas, and procedures.

---

[4]Even when taking into consideration that the Japanese sample included a large percentage of lessons on two-dimensional geometry, the percentage of problems per lesson of each level of procedural complexity remained relatively consistent (see figure 4.2).

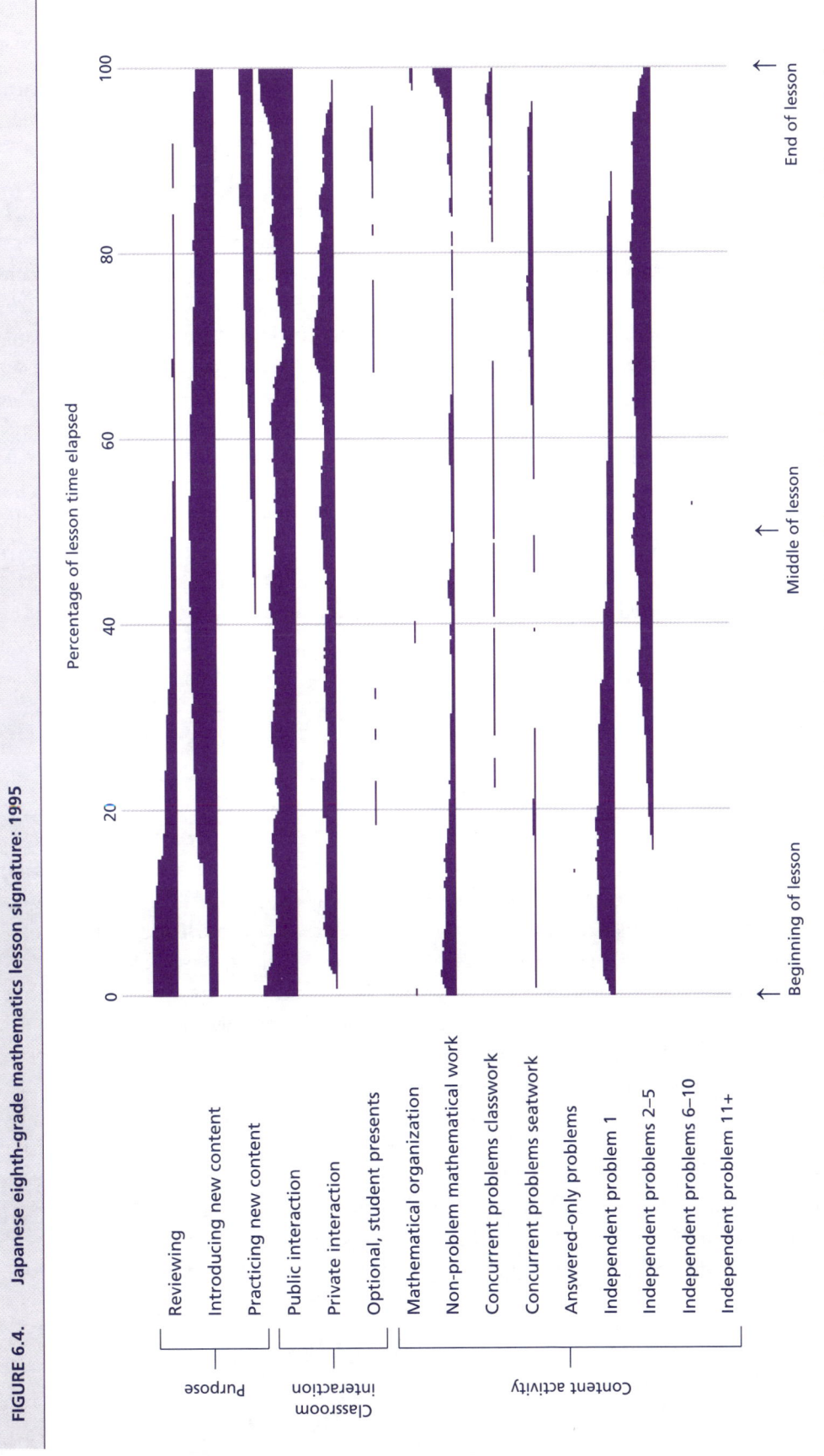

Chapter 6 | 137
Similarities and Differences in Eighth-Grade Mathematics Teaching Across Seven Countries

FIGURE 6.4. Japanese eighth-grade mathematics lesson signature: 1995

NOTE: The graph represents both the frequency of occurrence of a feature and the elapsing of time throughout a lesson. For each feature listed along the left side of the graph, the histogram (or bar) represents the percentage of eighth-grade mathematics lessons that exhibited the feature—the thicker the histogram, the larger the percentage of lessons that exhibited the feature. From left to right, the percentage of elapsed time in a lesson is marked along the bottom of the graph. The histogram increases by one pixel (or printable dot) for every 5 percent of lessons marked for a feature at any given moment during the lesson time, and disappears when fewer than 5 percent of lessons were marked (due to technological limitations). By following each histogram from left to right, one can get an idea of the percentage of lessons that included the feature as lesson time elapsed. A listing of the percentage of lessons that included each feature by the elapsing of time is included in appendix F. To create each histogram, each lesson was divided into 250 segments, each representing 0.4 percent of lesson time. The codes applied to each lesson at the start of each segment were tabulated, using weighted data, and reported as the percentage of lessons exhibiting each feature at particular moments in time.
SOURCE: U.S. Department of Education, National Center for Education Statistics, Third International Mathematics and Science Study (TIMSS), Video Study, 1999.

The lesson signature for the Netherlands

Purpose

As seen in other lesson signatures, a majority of eighth-grade Dutch mathematics lessons began with a review of previously learned content (64 percent), though a noticeable percentage of lessons began directly with the introduction of new content (29 percent, figure 6.5 and table F.5, appendix F). By the midpoint of the lesson, the percentage of lessons that were focused on review, introducing new content, or practicing new content, were relatively evenly divided (30, 34, and 29 percent, respectively, table F.5, appendix F). Overall, the majority of lesson time was spent on new content (either through the introduction of new content or its practice, 63 percent, figure 3.8) and nearly one-quarter of lessons were devoted entirely to review (24 percent, figure 3.9). The midpoint of the lesson is also the time when a majority of Dutch lessons moved into private interaction, wherein students worked individually or in small groups, and focused on sets of problems (concurrent problems) completed as seatwork. As suggested by country experts (see the country model, figure E.5, appendix E), Dutch students' first experiences with new concepts or procedures might come directly from the textbook, perhaps while working on homework assigned for the next day (on average, 10 problems started during the lesson were assigned to be completed as homework, table 3.8). In those instances, according to the experts, students usually are responsible for reading over the text sections as they work privately on completing problems. This view is consistent with analyses that show a high percentage of Dutch lessons were taught by mathematics teachers who cited the textbook as a major determinant of the lesson content (97 percent, table 2.6). Finally, as stated earlier, analyses revealed that it was not uncommon in Dutch eighth-grade mathematics lessons for the practice work begun by students in class to be continued at home. A relatively large portion of lesson time in comparison to the other countries—estimated to be 10 minutes of a 45-minute lesson, on average—was spent on problems that were assigned for future homework (table 3.8).

Classroom interaction

A majority of eighth-grade Dutch mathematics lessons began with whole-class, public discussion and this form of classroom interaction continued until almost the midpoint of the lesson (figure 6.5 and table F.5, appendix F). This period of time coincided with the time when a majority of the lessons were focused largely on a review of previously learned content. Based on an earlier analysis, Dutch lessons were estimated to spend among the largest amount of time on the public discussion of previously assigned homework in all the countries (16 minutes, on average, table 3.9). As observed in the videotapes, these discussions appeared to take place during the beginning of lessons, when a majority of lessons focused on review. As noted above, a majority of Dutch lessons moved into the introduction of new content or its practice at around 40 percent of the way through a lesson (figure 6.4 and table F.5, appendix F), which is close to the time in the lesson when a majority of Dutch lessons moved into individual or small-group work (private interaction). Among all the countries, eighth-grade Dutch mathematics lessons dedicated a greater percentage of lesson time to students working privately (55 percent, on average, table 3.6), largely individually rather than in small groups (on average, 90 percent of private interaction time per lesson was spent working individually, figure 3.10).

Content activities

The introduction and practice of new content in eighth-grade Dutch lessons coincided with an increasing percentage of lessons focused on individual student work on sets of problems (concurrent problems). As discussed above, and as suggested by country experts, eighth-grade Dutch students were observed to spend around half of the lesson time working individually. While working at their desks, eighth-grade Dutch students spent an average of 61 percent of lesson time on sets of problems (concurrent problems) rather than on independent problems (figure 3.4), among the highest percentages of all the countries. As displayed in the lesson signature (figure F.5), working on sets of problems commonly occurred during the second half of the lesson (table F.5, appendix F). Though all the countries utilized mathematics problems as the primary vehicle through which students came to acquire mathematical knowledge, eighth-grade Dutch lessons were found to devote a greater percentage of lesson time to working on problems than all of the countries except the United States (91 percent of lesson time, on average, figure 3.3). Mathematics problems in Dutch lessons were among the most frequent to be set up using a real-life connection (42 percent, figure 5.1) and to use calculators (91 percent of the lessons, figure 5.18). When experts reviewed the mathematics problems introduced into Dutch lessons for the level of procedural complexity—by looking at the number of steps it takes to solve a problem using common solution methods—69 percent of problems per lesson were deemed of low procedural complexity, 25 percent of moderate complexity, and 6 percent of high complexity (figure 4.1). Moreover, of all the problems introduced into the lessons, the majority of problems per lesson (57 percent) were posed with the apparent intent of using procedures, with another quarter posed with the apparent intent of making connections between ideas, facts, or procedures (24 percent, figure 5.8). When the problems introduced in the lessons were examined a second time for the processes made public while working through the problems, 36 percent of problems per lesson were found to be solved by explicitly using procedures and 22 percent were solved by actually making mathematical connections (figure 5.9). Of the 24 percent of problems per lesson that were posed to make mathematical connections, 37 percent per lesson were explicitly and publicly making the connections. Furthermore, examination of the sets of problems (concurrent problems) assigned to students to work on individually at their desks showed that almost three-quarters of the lesson time devoted to working privately focused on repeating procedures that had been demonstrated earlier in the lesson (74 percent, on average, figure 5.12).

Finally, homework appeared to play a role in the learning of content in Dutch lessons, as evidenced by the estimated time spent in discussion of previously assigned homework (16 minutes, on average, table 3.9) and the estimated percentage of lesson time spent on problems that were assigned for future homework (10 minutes, table 3.8). Country experts suggested that this relative emphasis on homework placed some responsibility on students for selecting what needed to be discussed at the beginning of the lesson and for working through the new content toward the end of the lesson (see figure E.5, appendix E).

All of these observations about eighth-grade Dutch mathematics lessons suggest that on average the introduction and practice of new content in eighth-grade Dutch lessons is often accomplished through students working individually on sets of mathematics problems that focus on using procedures. This is consistent with the country experts' assertion that Dutch students are expected to take responsibility for their own learning and are therefore given more independence or freedom to work on problems on their own or with others (figure E.5, appendix E).

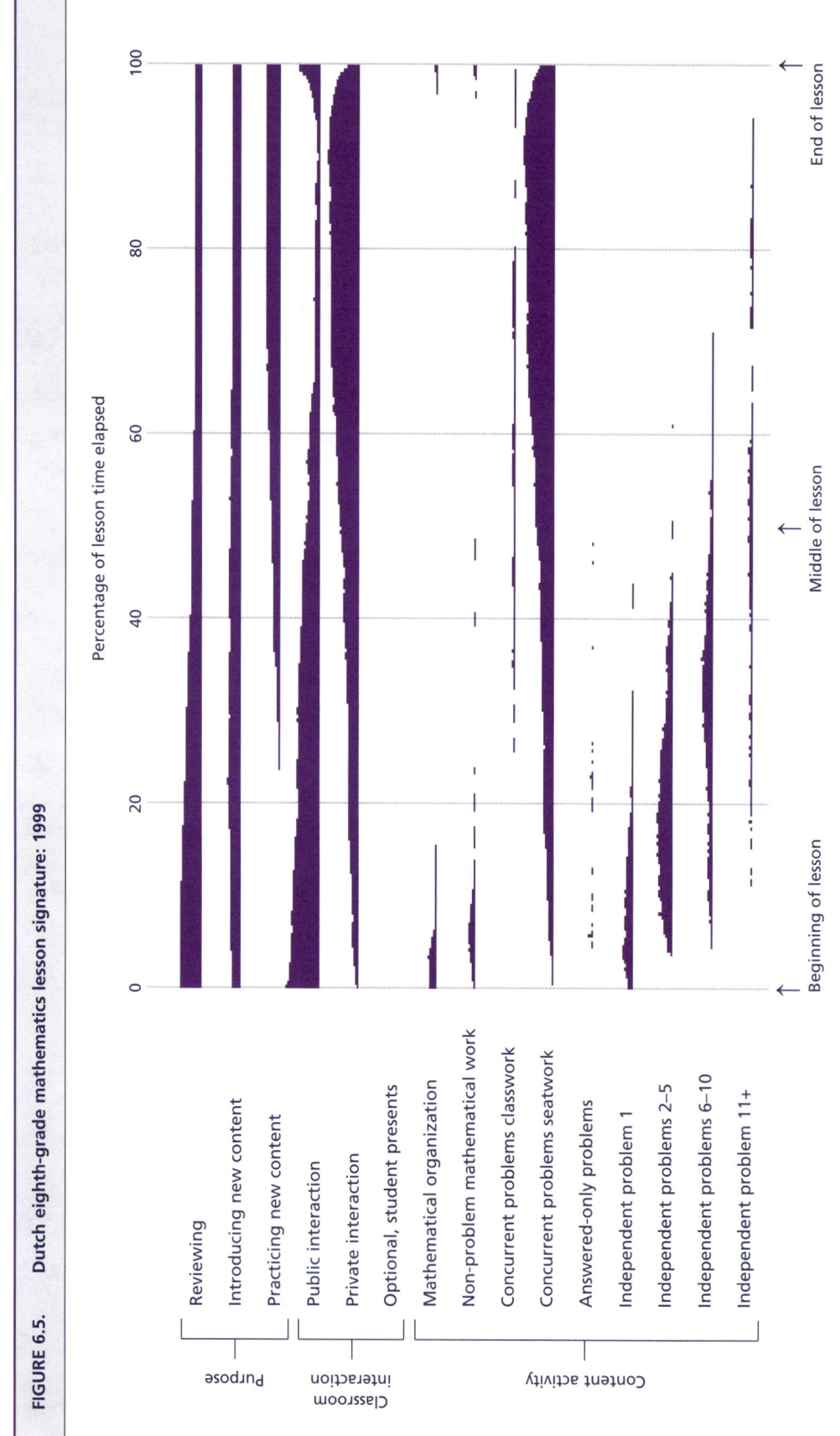

140 | Teaching Mathematics in Seven Countries
Results From the TIMSS 1999 Video Study

FIGURE 6.5. Dutch eighth-grade mathematics lesson signature: 1999

NOTE: The graph represents both the frequency of occurrence of a feature and the elapsing of time throughout a lesson. For each feature listed along the left side of the graph, the histogram (or bar) represents the percentage of eighth-grade mathematics lessons that exhibited the feature—the thicker the histogram, the larger the percentage of lessons that exhibited the feature. From left to right, the percentage of elapsed time in a lesson is marked along the bottom of the graph. The histogram increases by one pixel (or printable dot) for every 5 percent of lessons marked for a feature at any given moment during the lesson time, and disappears when fewer than 5 percent of lessons were marked (due to technological limitations). By following each histogram from left to right, one can get an idea of the percentage of lessons that included the feature as lesson time elapsed. A listing of the percentage of lessons that included each feature by the elapsing of time is included in appendix F. To create each histogram, each lesson was divided into 250 segments, each representing 0.4 percent of lesson time. The codes applied to each lesson at the start of each segment were tabulated, using weighted data, and reported as the percentage of lessons exhibiting each feature at particular moments in time.

SOURCE: U.S. Department of Education, National Center for Education Statistics, Third International Mathematics and Science Study (TIMSS), Video Study, 1999.

Chapter 6
Similarities and Differences in Eighth-Grade Mathematics Teaching Across Seven Countries

The lesson signature for Switzerland

Purpose

Seventy-one percent of eighth-grade Swiss mathematics lessons began with a review segment, with a majority of lessons maintaining that focus through the first 20 percent of lesson time (figure 6.6 and table F.6, appendix F). The review of previously introduced content constituted an average of 34 percent of lesson time (figure 3.8). As shown in the lesson signature, a majority of Swiss lessons shifted to the introduction and practice of new content by about one-third of the way through the lesson, with an increasing percentage of lessons focused on the practice of new content as lesson time elapsed (figure 6.6 and table F.6, appendix F). This observation is generally consistent with an earlier analysis that showed Swiss lessons devoted an average of 63 percent of lesson time to introducing and practicing new content (figure 3.8). During the time when a majority of Swiss lessons focused on the review of previously learned content, a majority of lessons were conducted through whole-class, public discussion (figure 6.6 and table F.6, appendix F). This follows the observations of country experts who suggested that during the review phase the teacher takes a leading role but involves students by asking questions and engaging in "interactive instruction" (see the hypothesized Swiss country model, figure E.6, appendix E).

Classroom interaction

Although whole-class, public interaction was common throughout most of the first half and at the very end of Swiss eighth-grade mathematics lessons, a majority of Swiss lessons also devoted lesson time to individual student or small-group work for a period of time during the second half of the lesson (figure 6.6 and table F.6, appendix F). Indeed, as reported in an earlier analysis, eighth-grade Swiss mathematics lessons used 44 percent of lesson time, on average, for individual student and small-group work, surpassed only by Dutch lessons (table 3.6). During this period of lesson time, students spent an average of three-quarters of the private time working individually, with the remaining one-quarter working in pairs or small groups (74 and 26 percent per lesson, figure 3.10). The time during which eighth-grade Swiss students worked individually or in small-group work largely coincided with the time during which, in a majority of lessons, students were asked to work on sets of problems (concurrent problems; figure 6.6 and table F.6, appendix F). According to Swiss country experts, working privately on problems is a common activity when students are practicing new content in order to become more efficient in executing solution procedures (see the hypothesized Swiss country model, figure E.6, appendix E).

Content activities

Although the content of some eighth-grade Swiss lessons was delivered by focusing on independent problems, the largest percentage of lessons utilized sets of problems (concurrent problems) as either whole-class or seatwork (figure 6.6 and table F.6, appendix F). Indeed, as revealed in earlier analyses, an average of 31 percent of lesson time in Switzerland was spent on independent problems and an average of 53 percent of lesson time was spent on sets of problems (concurrent problems; figure 3.4). According to Swiss country experts, two patterns of mathematics teaching were predicted to be observed in the videotaped lessons: one would focus on the introduction of new knowledge through a kind of Socratic dialogue between teacher and students (figure E.6, appendix E) and the second would focus largely on practicing content introduced in previous lessons (figure E.7, appendix E). Although analyses conducted for this study do not point to one or the other hypothesized model as being predominant in eighth-grade Swiss mathematics lessons, it seems relatively clear that, in the majority of lessons, the introduction and practice of new mathematics content was conveyed to students through working on sets of problems rather than through working on independent, individual problems (figure 6.6 and table F.6, appendix F). This is a feature of eighth-grade Dutch mathematics lessons as well, as pointed out earlier. Finally, a majority of problems introduced per lesson in eighth-grade Swiss lessons exhibited a low level of procedural complexity, meaning that these problems could be solved using four or fewer steps (figure 4.1). An additional 22 percent of problems per lesson were found to be of moderate complexity, and 12 percent of high complexity (figure 4.1).

Looking across the indicators of mathematics teaching in Switzerland suggests a mixed picture. From the lesson signature and other data, it appears that on average eighth-grade Swiss mathematics lessons devoted some time to review but spent the bulk of lesson time on the introduction and practice of new content. To convey to students the new content, Swiss lessons employed a mix of independent problems and sets of problems (concurrent problems), though a majority of lessons utilized sets of problems during most of the latter half of the lesson, which coincided with an increasing focus on the practice of new content. Among the participating countries, Switzerland is unique in that it operates under three separate educational systems, depending on the Canton (state) and the dominant language spoken in the area (i.e., French, Italian, or German) (Clausen, Reusser, and Klieme forthcoming). Assuming that the operation of these three systems results in different decisions being made about content and how it is taught in the classroom, this makes summarizing across the various indicators of mathematics teaching to find a "country-level" pattern challenging.

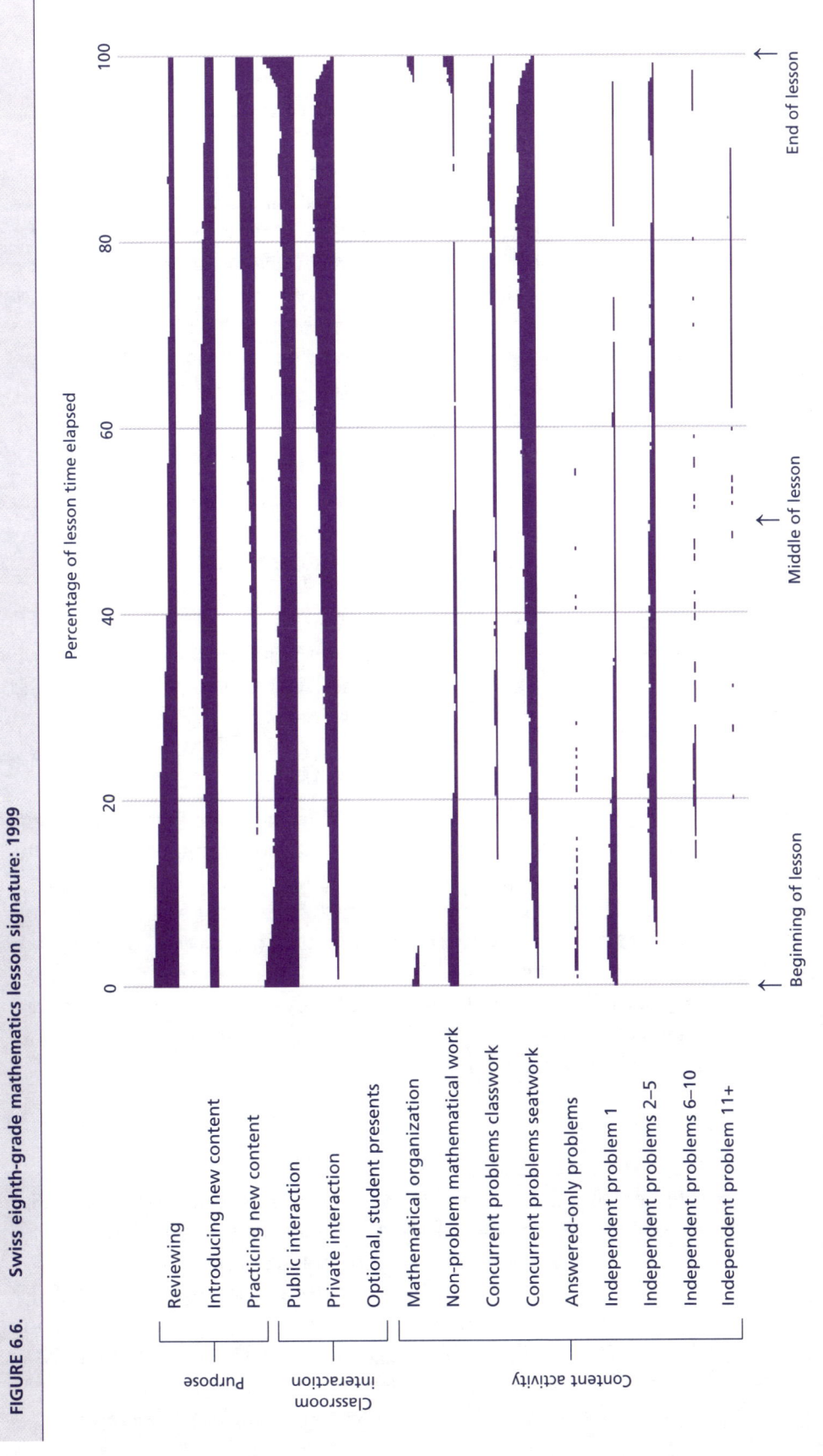

Chapter 6 | 143
Similarities and Differences in Eighth-Grade Mathematics Teaching Across Seven Countries

FIGURE 6.6. Swiss eighth-grade mathematics lesson signature: 1999

NOTE: The graph represents both the frequency of occurrence of a feature and the elapsing of time throughout a lesson. For each feature listed along the left side of the graph, the histogram (or bar) represents the percentage of eighth-grade mathematics lessons that exhibited the feature—the thicker the histogram, the larger the percentage of lessons that exhibited the feature. From left to right, the percentage of elapsed time in a lesson is marked along the bottom of the graph. The histogram increases by one pixel (or printable dot) for every 5 percent of lessons marked for a feature at any given moment during the lesson time, and disappears when fewer than 5 percent of lessons were marked (due to technological limitations). By following each histogram from left to right, one can get an idea of the percentage of lessons that included the feature as lesson time elapsed. A listing of the percentage of lessons that included each feature by the elapsing of time is included in appendix F. To create each histogram, each lesson was divided into 250 segments, each representing 0.4 percent of lesson time. The codes applied to each lesson at the start of each segment were tabulated, using weighted data, and reported as the percentage of lessons exhibiting each feature at particular moments in time.

SOURCE: U.S. Department of Education, National Center for Education Statistics, Third International Mathematics and Science Study (TIMSS), Video Study, 1999.

The lesson signature for the United States

Purpose

Through most of the first half of the lesson time in the United States, the majority of eighth-grade mathematics lessons focused on reviewing previously learned content (figure 6.7 and table F.7, appendix F). Indeed, an earlier analysis showed that, of the 94 percent of lessons that engaged students in review during some portion of the lesson (table 3.4), an average of 53 percent of lesson time was spent reviewing previously learned material, among the highest percentages of all the countries (figure 3.8). Moreover, teachers in 28 percent of eighth-grade U.S. mathematics lessons were found to spend the entire lesson on reviewing previously learned content, also one of the highest percentages among the countries examined (figure 3.9). The relative emphasis on review was predicted by country experts who were fairly detailed in their description of review and practice segments in comparison to the description of introducing new content (see the hypothesized United States country model, figure E.8, appendix E). For example, the hypothesized U.S. model created by country experts included three different goals for review: to assess or evaluate, to re-teach, and to "warm-up" in preparation for the lesson.

Around half way through the lesson, a majority of eighth-grade U.S. mathematics lessons shifted focus to the introduction and practice of new content (figure 6.7 and table F.7, appendix F). Nonetheless, averaging across all the eighth-grade mathematics lessons, the United States was among the countries with the smallest percentage of lesson time devoted to introducing and practicing new content (48 percent, figure 3.8). Although in some of the countries there was a detectable difference in the emphasis placed within the lessons on either reviewing previously learned content or introducing and practicing new content, there was no such difference found in the United States (figure 3.8).

Classroom interaction

On average, 67 percent of eighth-grade mathematics lesson time in the United States was spent in whole-class, public interaction (table 3.6). This pattern was relatively prominent throughout most of the lesson, as seen in the lesson signature (figure 6.7 and table F.7, appendix F). The United States was one of the few countries in which some lessons began with students working on a set of problems as seatwork (21 percent, table F.7, appendix F). This may be consistent with what the country experts described in the hypothesized country model as the conducting of a "warm-up" activity, which is reportedly designed to secure and activate old knowledge (figure E.8, appendix E). One way in which the lesson might conclude, suggests the hypothesized model, is for students to practice new material while working on their own.

Content activities

As with the other countries, the delivery of content in eighth-grade U.S. mathematics lessons was accomplished primarily by working through problems. As noted above, a majority of U.S. eighth-grade mathematics lessons focused on the introduction and practice of new content beginning in the second half of the lesson. Though U.S. eighth-grade mathematics teachers appeared to engage students with both independent and sets of problems, a majority of U.S. lessons utilized independent problems to convey mathematical content during the middle portion of the lesson (figure 6.7 and table F.7, appendix F). This is consistent with an earlier analysis that showed 51 percent of U.S. mathematics lesson time devoted to working on independent problems, on average (figure 3.4). Moreover, another analysis showed that, on average, an eighth-grade U.S. mathematics lesson included 10 independent problems, among the highest frequency of all the countries (table 3.3). When eighth-grade students worked privately—as when they were assigned sets of problems (concurrent problems) to be completed in seatwork—students usually worked individually (80 percent of private interaction time per lesson, figure 3.10) and usually spent their time repeating procedures that were introduced earlier in the lesson (75 percent of private time per lesson; figure 5.13).

When taking into consideration all the problems presented in the U.S. lessons, 69 percent of the problems per lesson were found to be posed with the apparent intent of using procedures—problems that are typically solved by applying a procedure or set of procedures—a higher percentage than problems that were posed with the apparent intent of making connections between ideas, facts, or procedures, or problems that were posed with the apparent intent of eliciting a mathematical convention or concept (stating concepts; figure 5.8). When the problems introduced in the lesson were examined a second time for processes made public while working through the problems, 91 percent of the problems per lesson in the United States were found to have been solved by giving results only without discussion of how the answer was obtained or by focusing on the procedures necessary to solve the problem (figure 5.9). Moreover, when the 17 percent of problems per lesson that were posed to make mathematical connections were followed through to see whether the connections were stated or discussed publicly, less than one percent per lesson were solved by explicitly and publicly making the connections (figure 5.12). Finally, an expert review of the mathematics problems introduced into U.S. lessons revealed that 67 percent of problems per lesson were deemed of low procedural complexity—based on the number of steps it took to solve a problem using common solution methods (figure 4.1). An additional 27 percent of problems per lesson were found to be of moderate procedural complexity, and 6 percent of high complexity.

All of these observations suggest that on average U.S. eighth-grade mathematics lessons were conducted largely through whole-class, public discussions that focused students' attention on both previously learned and new content by working on multiple independent problems, supplemented by practice on the occasional set of problems, with the goal of learning and using procedures.

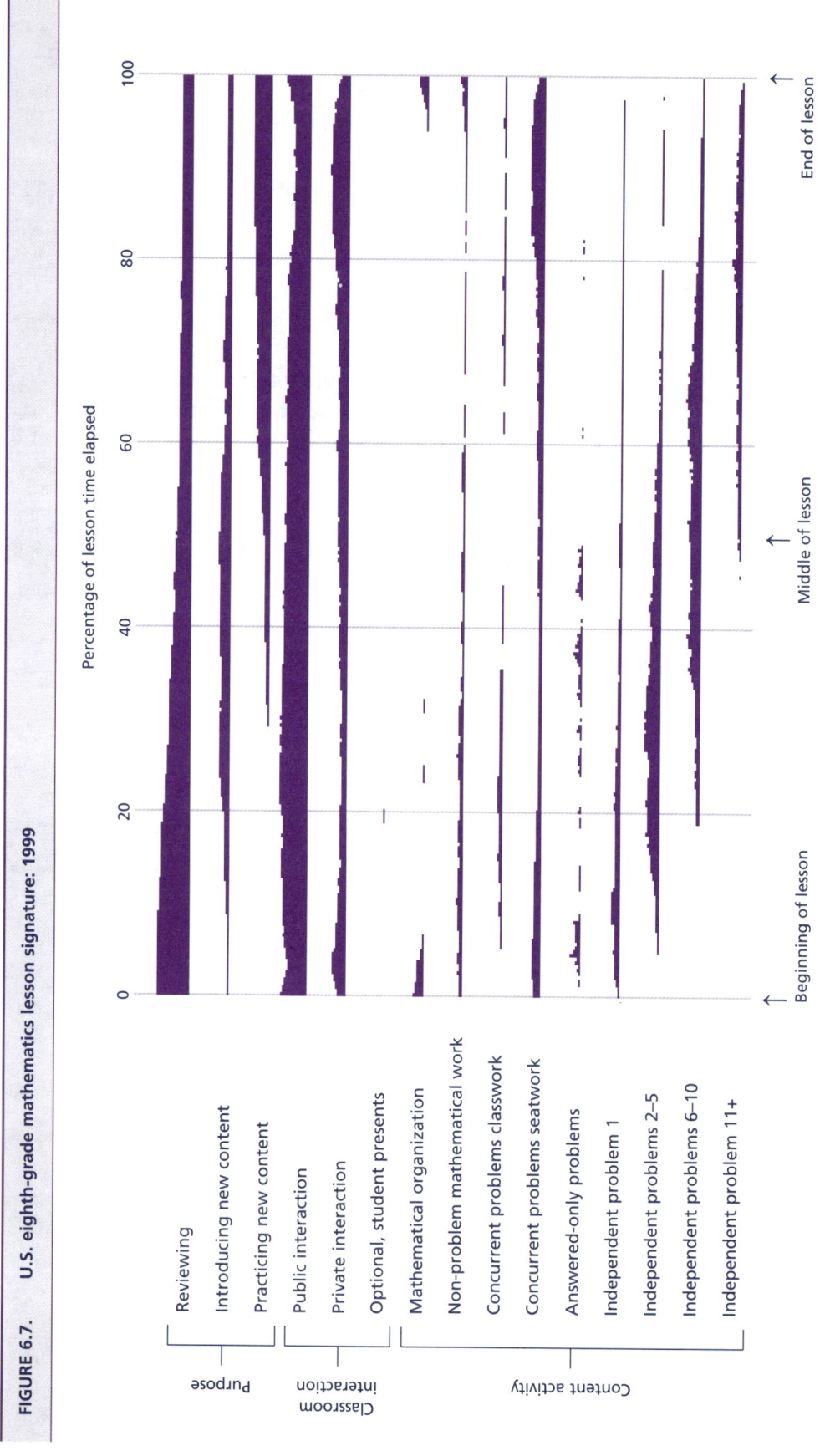

146 | Teaching Mathematics in Seven Countries
Results From the TIMSS 1999 Video Study

FIGURE 6.7. U.S. eighth-grade mathematics lesson signature: 1999

NOTE: The graph represents both the frequency of occurrence of a feature and the elapsing of time throughout a lesson. For each feature listed along the left side of the graph, the histogram (or bar) represents the percentage of eighth-grade mathematics lessons that exhibited the feature—the thicker the histogram, the larger the percentage of lessons that exhibited the feature. From left to right, the percentage of elapsed time in a lesson is marked along the bottom of the graph. The histogram increases by one pixel (or printable dot) for every 5 percent of lessons marked for a feature at any given moment during the lesson time, and disappears when fewer than 5 percent of lessons were marked (due to technological limitations). By following each histogram from left to right, one can get an idea of the percentage of lessons that included the feature as lesson time elapsed. A listing of the percentage of lessons that included each feature by the elapsing of time is included in appendix F. To create each histogram, each lesson was divided into 250 segments, each representing 0.4 percent of lesson time. The codes applied to each lesson at the start of each segment were tabulated, using weighted data, and reported as the percentage of lessons exhibiting each feature at particular moments in time.

SOURCE: U.S. Department of Education, National Center for Education Statistics, Third International Mathematics and Science Study (TIMSS), Video Study, 1999.

Summary

Information about eighth-grade mathematics teaching displayed in the lesson signatures and the accompanying descriptions complements the information presented in chapters 2 to 5 in several ways. First, the time during the lesson that particular lesson features were evident reveals the flow of lessons in ways that were not apparent from comparing the occurrence of individual features. Second, the co-occurrence of particular features during the lesson suggested interpretations about the nature of individual features. For example, non-problem segments that occurred at the beginning of the lesson along with a high percentage of public interaction and review suggested that these segments involved a whole-class discussion or brief presentation on material previously learned. Third, by invoking information from country experts and contained in the hypothesized country models, it was possible to speculate about the meaning of particular patterns found in the signatures. In general, the country models helped to put back together individual features separated for coding and analysis.

A summary of impressions about individual countries gained from viewing the lesson signatures includes the following. As noted earlier in this report (figure 3.8), the emphasis on introducing new material in Japan, on practicing new material in Hong Kong SAR, and on review in the Czech Republic and the United States all were reinforced in the lesson signatures. The emphasis in the Netherlands on private work also was supported.

Another set of impressions concerns the issue of convergence versus variability across mathematics lessons within countries. The height of the histograms, or the width of the bands, in the lesson signature graphs indicates the extent to which eighth-grade mathematics lessons for a particular country displayed the same profile across time. Lessons in Hong Kong SAR and Japan showed some convergence along the purpose dimensions, and lessons in the Netherlands showed particular convergence in a mid-lesson shift from public to private interaction. Other countries showed less convergence, as indicated by much overlap among variables within a dimension and few definitive peaks in the histograms.

Variability among lessons within countries might come from several sources. One source is individual differences among teachers. Different teachers might organize and implement lessons in different ways. Another source is the more systematic differences that could result from several different methods of teaching co-existing within a single country. One of the hypotheses arising from the TIMSS 1995 Video Study was that lessons within countries show considerable similarity compared with lessons across countries (Stigler and Hiebert 1999). These similarities might result from cultural "scripts" of teaching that become widely shared within a country.

Although the data presented in this report are, in general, consistent with this hypothesis, the variability within some countries suggests a more complicated picture. Switzerland is an instructive case. The lesson signature for Switzerland (figure 6.6) showed a lack of convergence along several dimensions. There appeared to be no clear consistency among lessons with regard to when during the lesson particular features were evident. Concurrent research in Switzerland suggests that this variability might be explained by the different language areas within the country (Clausen, Reusser, and Klieme forthcoming) and by educational reform activity, currently underway, that has yielded two different methods of teaching mathematics (Reusser et al. forthcoming). These results suggest it is useful to search the lesson signatures for indicators of variability within countries as well as for points of convergence.

The Roles Played by Individual Lesson Features Within Different Teaching Systems

The fact that lessons within each country can accumulate to display a characteristic pattern means that the same basic ingredients of lessons can be arranged to yield recognizable systems of teaching (Stigler and Hiebert 1999). A system of teaching can be thought of as a recurring similarity in the way in which the basic lesson ingredients interact. These interactions change during a lesson as new ingredients are introduced, ingredients in-use disappear, and the emphasis placed on particular sets of ingredients waxes and wanes. A partial picture of systems, based on a small set of variables, is portrayed by the lesson signatures. As noted above, there are some conditions under which more than one system of teaching might operate within a given country.

From a system perspective, the meaning of each ingredient depends on its role in the system: when it occurs, the other features that co-occur, and the function it is serving at the time. The same ingredients can mean different things to the students (and the teacher) within different systems of teaching. The consequence is that individual ingredients or features can provide more information about teaching than is obtained through comparing them one at a time outside of the systems in which they are functioning. What is needed is a description of the system(s) of teaching in each country and then an analysis of the roles played by individual features within each system.

A full system analysis of eighth-grade mathematics teaching is beyond the scope of this study, in part because it would require a richer database than is available here (e.g., more than one lesson per teacher). But the lesson signatures presented earlier provide a preview, suggested by the data available, of country-based systems of teaching. This section extends the system analysis by considering two examples of the way in which similar ingredients can play quite different roles within different systems of teaching.

As a first example, consider the different roles that private interaction can play in eighth-grade mathematics lessons. Private interaction can occur at different points in the lesson (see the lesson signatures), and it can be used to engage students in different kinds of work. As stated earlier, eighth-grade students in Japan engaged in activities other than repeating procedures during a greater percentage of private work time than students in the other countries (figure 5.13). In addition, by viewing the co-occurrence of private interaction with lesson segments of different purposes, the lesson signatures suggest that some private work time might be used to introduce new content (Japan, the Netherlands, and Switzerland, figures 6.4, 6.5, and 6.6), to practice procedures introduced during the lesson (Australia, the Czech Republic, the Netherlands, and the United States; figures 6.1, 6.2, 6.5 and 6.7), and to review procedures and definitions already learned (the Czech Republic and the United States; figures 6.2 and 6.7). The different purposes indicate that, by definition, private work time can play different roles and provide different kinds of learning opportunities for students.

As a second example of the different meanings that can be associated with the same lesson ingredient, consider reviewing in the Czech Republic and the United States. Eighth-grade mathematics lessons in the Czech Republic devoted a greater percentage of time, on average, to review (58 percent), than any of the other countries except the United States (53 percent; figure 3.8). It appears, however, that it would be a mistake to assume that this shared emphasis on reviewing results in a similar experience for Czech and U.S. students. As described in the following paragraphs, the lesson signatures for each country signal one of the differences and hint at its meaning.

The review segments occurred near the beginning of lessons in both countries (figures 6.2, 6.7). But during the review segments of the Czech mathematics lessons, a mixed form of interaction could occur. When viewing the videotapes, and by studying the hypothesized Czech country model (see appendix E), it is apparent that this mixed form often occurred when one or two students were called to the front of the room to be publicly "graded." The teacher assigned a review problem for the student(s) to work on the chalkboard while the rest of the students attended to the dialogue between the teacher and the student(s) being graded or, sometimes, were given a choice to either watch and listen or to work on their own. After the student(s) finished the problem on the chalkboard, the teacher asked the student(s) questions about the work and then announced the grade.

Segments like the Czech grading were not seen in the United States lessons. Impressions from viewing the videotapes and information in the hypothesized country models (appendix E) indicated that reviewing in the United States usually occurred through a whole-class discussion, with the teacher answering questions and working through problems at the chalkboard requested by the students, or through students working individually on a set of "warm-up" problems. These different kinds of segments in the Czech Republic and the United States were all marked review, but it seems they could provide different learning experiences for the students.

The examples indicate that differences in teaching exist among the countries in this study, even along individual variables that might show similar frequencies of occurrence. To examine fully these more subtle differences, a combination of quantitative and qualitative analyses would be needed to identify patterns or systems of teaching and then to analyze the roles played within these systems by individual features of teaching.

Conclusions

There are no simple or easy stories to tell about eighth-grade mathematics teaching from the TIMSS 1999 Video Study results. More than anything, the findings of this study expand the discussion of teaching by underscoring its complexity.

One thing is clear however: the countries that show high levels of achievement on TIMSS do not all use teaching methods that combine and emphasize features in the same way. Different methods of mathematics teaching can be associated with high scores on international achievement tests. Eighth-grade Japanese students often perform well in mathematics (Gonzales et al. 2000), and Japanese eighth-grade mathematics teaching contains a number of distinctive features. Nonetheless, it appears that these features are not a necessary condition for high achievement in other countries. Teachers in Hong Kong SAR, the other participating country with a TIMSS mathematics score as high as Japan, used methods of teaching that contained a number of features different from Japan, while teachers in the other high-achieving countries employed still different features.

The comparison between Japan and Hong Kong SAR is especially instructive because they were the two highest achieving countries in the study (table 6.1). In both countries, 76 percent of lesson time, on average, was spent working with new content and 24 percent of lesson time was spent reviewing previous content (figure 3.8). The new content introduced in mathematics lessons in these countries was worked with in different ways however. In Japanese lessons, more time (than in all the other countries) was devoted to introducing the new content and in Hong Kong SAR more time (than in the Czech Republic, Japan, and Switzerland) was devoted to practicing the new content (figure 3.8). Consistent with this emphasis, a larger percentage of mathematics problems in Japanese eighth-grade mathematics lessons (than in all the other countries except the Netherlands) were presented with the apparent intent of asking students to make mathematical connections, and a larger percentage of mathematics problems in Hong Kong SAR lessons (than in all the other countries except the Czech Republic) were presented with the apparent intent of asking students to use procedures (figure 5.8). These different emphases are reinforced by recalling that a larger percentage of private work time in Hong Kong SAR lessons (along with those in the Czech Republic) was devoted to repeating procedures already learned than in Japanese (and Swiss) lessons (figure 5.13). Given that students in both Japan and Hong Kong SAR have performed well on international achievement tests such as TIMSS, it is interesting that their instructional practices (summarized in table 6.2) lie on the opposite ends of these dimensions.

TABLE 6.2. Similarities and differences between eighth-grade mathematics lessons in Japan and Hong Kong SAR on selected variables: 1995 and 1999

Lesson variable	Japan[1]	Hong Kong SAR
Reviewing[2] (figure 3.8)	24 percent of lesson time	24 percent of lesson time
New content (figure 3.8)	76 percent of lesson time	76 percent of lesson time
Introducing new content[3]	60 percent of lesson time	39 percent of lesson time
Practicing new content[4]	16 percent of lesson time	37 percent of lesson time
Problems[5,6] (as stated) (figure 5.8)	Making connections (54 percent of problems)	Using procedures (84 percent of problems)
Private work activity[7,8] (figure 5.13)	Something other than practicing procedures or mix (65 percent of work time)	Practicing procedures (81 percent of work time)

[1] Japanese mathematics data were collected in 1995.
[2] Reviewing: No differences detected.
[3] Introducing new content: JP>HK.
[4] Practicing new content: HK>JP.
[5] Making connections: JP>HK.
[6] Using procedures: HK>JP.
[7] Percent of private time devoted to something other than practicing procedures or mix: JP>HK.
[8] Percent of private time devoted to practicing procedures: HK>JP.
SOURCE: U.S. Department of Education, National Center for Education Statistics, Third International Mathematics and Science Study (TIMSS), Video Study, 1999.

If the learning goal for students is high performance on assessments of mathematics, the findings of this study suggest that there is no single method that mathematics teachers in relatively high-achieving countries use to achieve that goal. Different methods of mathematics teaching were found in different high-achieving countries. This conclusion suggests that informed choices of which teaching methods to use will require more detailed descriptions of learning goals than simply high performance on international tests. A particular country might have specific learning goals that are highly valued (see chapter 2, figure 2.1 and table 2.5) and for which particular methods of teaching may be better aligned than others. The results of this study make it clear that an international comparison of teaching, even among mostly high-achieving countries, cannot, by itself, yield a clear answer to the question of which method of mathematics teaching may be best to implement in a given country.

At the same time, the results of the study suggest that there are many similarities across countries, especially in the basic ingredients used to construct eighth-grade mathematics lessons. It is likely that many teachers in each country are familiar with these ingredients. All participating countries, for example, devoted at least 80 percent of lesson time, on average, to solving mathematics problems (figure 3.3) and all countries devoted some lesson time, on average, to presenting new content (figure 3.8) However, mathematics teachers in the different countries used these ingredients with different emphases or arranged them in different ways.

Interpreting the results from this study requires a thoughtful and analytic approach, including follow-up analyses and research that can more precisely examine the possible effects that particular methods or approaches may have on student learning. Through these kinds of activities, the ultimate aim of a study such as this can be realized: a deeper understanding of classroom mathematics teaching and a deeper understanding of how teaching methods can be increasingly aligned with learning goals for students.

REFERENCES

Anderson-Levitt, K.M. (2002). Teaching Culture as National and Transnational: A Response to *Teachers' Work*. *Educational Researcher*, 31(3), 19–21.

Bailey, B.J.R. (1977). Tables of the Bonferroni *t* Statistic. *Journal of the American Statistical Association*, 72, 469–478.

Bakeman, R., and Gottman, J.M. (1997). *Observing Interaction: An Introduction to Sequential Analysis*. Second Edition. Cambridge: Cambridge University Press.

Ball, D.L. (1993). Halves, Pieces, and Twoths: Constructing and Using Representational Contexts in Teaching Fractions. In T.P. Carpenter, E. Fennema, and T.A. Romberg (Eds.), *Rational Numbers: An Integration of Research* (pp. 157–195). Hillsdale, NJ: Erlbaum.

Beaton, A., Mullis, I.V.S., Martin, M.O., Gonzalez, E.J., Kelly, D.L., and Smith, T.A. (1996). *Mathematics Achievement in the Middle School Years: IEA's Third International Mathematics and Science Study*. Chestnut Hill, MA: Boston College.

Bishop, A.J., Clements, J., Keitel, C., Kilpatrick, J., and Laborde, C. (Eds.). (1996). *International Handbook of Mathematics Education*. Dordrecht, The Netherlands: Kluwer.

Brophy, J. (1999). Teaching (Education Practices Series No. 1). Geneva: International Bureau of Education. Available online at *http://www.ibe.unesco.org*.

Brophy, J.E., and Good, T.L. (1986). Teacher Behavior and Student Achievement. In M.C. Wittrock (Ed.), *Handbook of Research on Teaching* (3rd ed., pp. 328–375). New York: Macmillan.

Brownell, W.A. (1935). Psychological Considerations in the Learning and Teaching of Arithmetic. In W.D. Reeve (Ed.), *The Teaching of Arithmetic: Tenth Yearbook of the National Council of Teachers of Mathematics* (pp. 1–31). New York: Teachers College, Columbia University.

Bunyi, G. (1997). Multilingualism and Discourse in Primary School Mathematics in Kenya. *Language, Culture, and Curriculum*, 10 (1), 52–65.

Burkhardt, H. (1981). *The Real World and Mathematics*. London: Blackie.

Carver, C.S., and Scheier, M.F. (1981). *Attention and Self-Regulation: A Control-Theory Approach to Human Behavior*. New York: Springer-Verlag.

Cazden, C. (1988). *Classroom Discourse: The Language of Teaching and Learning*. Portsmouth, NH: Heinemann.

Clarke, D.J. (2003). International Comparative Studies in Mathematics Education. In A.J. Bishop, M.A. Clements, C. Keitel, J. Kilpatrick, and F.K.S. Leung (Eds.), *Second International Handbook of Mathematics Education,* (pp. 145–186). Dordrecht: Kluwer Academic Publishers.

Clausen, M., Reusser, K., and Klieme, E. (forthcoming). Unterrichtsqualität auf der Basis Hochinferenter Unterrichtsbeurteilungen: Ein Vergleich Zwischen Deutschland und der Deutschsprachigen Schweiz. *Unterrichtswissenschaft,* 31 (1).

Davis, P.J., and Hersh, R. (1981). *The Mathematical Experience.* Boston: Houghton Mifflin.

De Brigard, E. (1995). The History of Ethnographic Film. In P. Hockings (Ed.), *Principles of Visual Anthropology* (2nd ed., pp. 13–43). New York: Mouton de Gruyter.

Ember, C.R., and Ember, M. (1998). Cross-Cultural Research. In H.R. Bernard (Ed.), *Handbook of Methods in Cultural Anthropology* (pp. 647–687). Walnut Creek: Sage Publications.

English, L.D. (Ed.). (2002). *Handbook of International Research in Mathematics Education.* Mahwah, NJ: Erlbaum.

Fennema, E., and Franke, M.L. (1992). Teachers' Knowledge and Its Impact. In D. Grouws (Ed.), *Handbook of Research on Mathematics Teaching and Learning* (pp. 147–164). New York: Macmillan.

Fey, J.T. (1979). Mathematics Teaching Today: Perspective From Three National Surveys. *Arithmetic Teacher,* 27 (2), 10–14.

Fey, J.T., and Hirsch, C.R. (Eds.). (1992). *Calculators in Mathematics Education: 1992 Yearbook of the National Council of Teachers of Mathematics.* Reston, VA: National Council of Teachers of Mathematics.

Fisher, R.A. (1951). *The Design of Experiments.* Edinburgh: Oliver and Boyd.

Floden, R.E. (2001). Research on Effects of Teaching: A Continuing Model for Research on Teaching. In V. Richardson (Ed.), *Handbook of Research on Teaching* (4th ed., pp. 3–16). Washington, DC: American Educational Research Association.

Gage, N.L. (1978). *The Scientific Basis of the Art of Teaching.* New York: Teachers College Press, Columbia University.

Geertz, C. (1984). "From the Native's Point of View": On the Nature of Anthropological Understanding. In R.A. Shweder and R. LeVine (Eds.), *Culture Theory: Essays on Mind, Self, and Emotion* (pp. 123–136). Cambridge: Cambridge University Press.

Goldenberg, C.N. (1992/93). Instructional Conversations: Promoting Comprehension Through Discussion. *The Reading Teacher,* 46, 316–326.

Gonzales, P., Calsyn, C., Jocelyn, L., Mak, K., Kastberg, D., Arafeh, S., Williams, T., and Tsen, W. (2000). *Pursuing Excellence: Comparisons of International Eighth-Grade Mathematics and Science Achievement From a U.S. Perspective, 1995 and 1999.* NCES 2001-028. U.S. Department of Education. Washington, DC: National Center for Education Statistics.

Goodlad, J. (1984). *A Place Called School.* New York: McGraw-Hill.

Grouws, D.A. (Ed.). (1992). *Handbook of Research on Mathematics Teaching and Learning.* New York: Macmillan.

Hatano, G. (1988). Social and Motivational Bases for Mathematical Understanding. In G.B. Saxe and M. Gearhart (Eds.), *Children's Mathematics* (pp. 55–70). San Francisco: Jossey-Bass.

Hiebert, J. (Ed.). (1986). *Conceptual and Procedural Knowledge: The Case of Mathematics.* Mahwah, NJ: Erlbaum.

Hiebert, J. (1999). Relationships Between Research and the NCTM Standards. *Journal for Research in Mathematics Education,* 30, 3–19.

Hiebert, J., Carpenter, T.P., Fennema, E., Fuson, K.C., Human, P., Murray, H., Olivier, A., and Wearne, D. (1996). Problem Solving as a Basis for Reform in Curriculum and Instruction: The Case of Mathematics. *Educational Researcher,* 25 (4), 12–21.

Hiebert, J., Carpenter, T.P., Fennema, E., Fuson, K.C., Wearne, D., Murray, H., Olivier, A., and Human, P. (1997). *Making Sense: Teaching and Learning Mathematics With Understanding.* Portsmouth, NH: Heinemann.

Hiebert, J., and Wearne, D. (1993). Instructional Tasks, Classroom Discourse, and Students' Learning in Second-Grade Arithmetic. *American Educational Research Journal,* 30, 393–425.

Hoetker, J., and Ahlbrand, W. (1969). The Persistence of Recitation. *American Educational Research Journal,* 6, 145–167.

Jacobs, J., Garnier, H., Gallimore, R., Hollingsworth, H., Givvin, K.B., Rust, K., Kawanaka, T., Smith, M., Wearne, D., Manaster, A., Etterbeek, W., Hiebert, J., and Stigler, J.W. (forthcoming). *TIMSS 1999 Video Study Technical Report: Volume 1: Mathematics Study.* U.S. Department of Education. Washington, DC: National Center for Education Statistics.

Jacobs, J.K., Kawanaka, T., and Stigler, J.W. (1999). Integrating Qualitative and Quantitative Approaches to the Analysis of Video Data on Classroom Teaching. *International Journal of Educational Research,* 31, 717–724.

Kaput, J.J. (1992). Technology and Mathematics Education. In D.A. Grouws (Ed.), *Handbook of Research on Mathematics Teaching and Learning* (pp. 515–556). New York: Macmillan.

Knapp, M., and Shields, P. (1990). Reconceiving Academic Instruction for the Children of Poverty. *Phi Delta Kappan,* 71, 753–758.

Knoll, S., and Stigler, J.W. (1999). Management and Analysis of Large-Scale Video Surveys Using the Software vPrism™. *International Journal of Educational Research,* 31, 725–734.

Lampert, M. (2001). *Teaching Problems and the Problems of Teaching.* New Haven, CT: Yale University Press.

Leinhardt, G. (1986). Expertise in Math Teaching. *Educational Leadership,* 43 (7), 28–33.

Lesh, R., and Lamon, S.J. (Eds.). (1992). *Assessment of Authentic Performance in School Mathematics.* Washington, DC: American Association for the Advancement of Science.

LeTendre, G., Baker, D., Akiba, M., Goesling, B., and Wiseman, A. (2001). Teachers' Work: Institutional Isomorphism and Cultural Variation in the U.S., Germany, and Japan. *Educational Researcher,* 30 (6), 3–15.

Leung, F.K.S. (1995). The Mathematics Classroom in Beijing, Hong Kong, and London. *Educational Studies in Mathematics,* 29, 297–325.

Manaster, A.B. (1998). Some Characteristics of Eighth Grade Mathematics Classes in the TIMSS Videotape Study. *American Mathematical Monthly,* 105, 793–805.

Martin, M.O., Gregory, K.D., and Stemler, S.E. (2000). *TIMSS 1999 Technical Report.* Chestnut Hill, MA: Boston College.

Martin, M.O., Mullis, I.V.S., Gonzalez, E.J., Gregory, K.D., Smith, T.A., Chrostowski, S.J., Garden, R.A., and O'Connor, K.M. (2000). *TIMSS 1999 International Science Report: Findings From IEA's Repeat of the Third International Mathematics and Science Study at the Eighth Grade.* Chestnut Hill, MA: Boston College.

McKnight, C.C., Crosswhite, F.J., Dossey, J.A., Kifer, E., Swafford, J.O., Travers, K.J., and Cooney, T.J. (1987). *The Underachieving Curriculum: Assessing U.S. Schools Mathematics From an International Perspective.* Champaign, IL: Stipes.

Mullis, I.V.S, Jones, C., and Garden, R.A. (1996). Training for Free Response Scoring and Administration of Performance Assessment. In M.O. Martin and D.L. Kelly (Eds.), *Third International Mathematics and Science Study Technical Report, Volume 1: Design and Development.* Chestnut Hill, MA: Boston College.

Mullis, I.V.S., and Martin, M.O. (1998). Item Analysis and Review. In M.O. Martin and D.L. Kelly (Eds.), *Third International Mathematics and Science Study Technical Report, Volume II: Implementation and Analysis, Primary and Middle School Years (Population 1 and Population 2).* Chestnut Hill, MA: Boston College.

Mullis, I.V.S., Martin, M.O., Gonzalez, E.J., Gregory, K.D., Garden, R.A., O'Connor, K.M., Chrostowski, S.J., and Smith, T.A. (2000). *TIMSS 1999 International Mathematics Report: Findings From IEA's Repeat of the Third International Mathematics and Science Study at the Eighth Grade.* Chestnut Hill, MA: Boston College.

National Council of Teachers of Mathematics. (1989). *Curriculum and Evaluation Standards for School Mathematics.* Reston, VA: National Council of Teachers of Mathematics.

National Council of Teachers of Mathematics. (2000). *Principles and Standards for School Mathematics.* Reston, VA: National Council of Teachers of Mathematics.

National Research Council. (1999). *How People Learn: Brain, Mind, Experience, and School.* J.D. Bransford, A.L. Brown, and R.R. Cocking (Eds.). Committee on Developments in the Science of Learning, Commission on Behavioral and Social Sciences and Education. Washington, DC: National Academy Press.

National Research Council. (2001a). *Adding it up: Helping Children Learn Mathematics.* J. Kilpatrick, J. Swafford, and B. Findell (Eds.). Mathematics Learning Study Committee, Center for Education, Division of Behavioral and Social Sciences and Education. Washington, DC: National Academy Press.

National Research Council. (2001b). *The Power of Video Technology in International Comparative Research in Education.* Washington, DC: National Academy Press.

Neubrand, J. (forthcoming). Characteristics of Problems in the Lessons of the TIMSS Video Study. In B. Kaur and B.H. Yeap (Eds.), *TIMSS and Comparative Studies in Mathematics Education: An International Perspective* (pp. 72–80). Singapore: National Institute of Education.

Nisbett, R.E., and Ross, L. (1980). *Human Inference: Strategies and Shortcomings of Social Judgment.* Englewood Cliffs, NJ: Prentice-Hall.

Organization for Economic Co-operation and Development. (2001). *Knowledge and Skills for Life: First Results From the OECD Programme for International Student Assessment* (PISA) 2000. Organization for Economic Co-operation and Development: Paris.

Prawat, R.S. (1991). The Value of Ideas: The Immersion Approach to the Development of Thinking. *Educational Researcher,* 20 (2), 3–10.

Reusser, K., Pauli, C., Waldis, M., and Grob, U. (forthcoming). Erweiterte Lernformen – auf dem Weg zu Einem Adaptiven Mathematikunterricht in der Deutschscweiz. *Unterrichtswissenschaft,* 31 (3).

Richardson, V. (Ed.). (2001). *Handbook of Research on Teaching* (4th ed.). Washington, DC: American Educational Research Association.

Robitaille, D.F. (1995). *Mathematics Textbooks: A Comparative Study of Grade 8 Texts.* Vancouver, Canada: Pacific Education Press.

Rosenshine, B., and Stevens, R. (1986). Teaching Functions. In M. Wittrock (Ed.), *Handbook of Research on Teaching* (3rd ed., pp. 376–391). New York: Macmillan.

Ruthven, K. (1996). Calculators in Mathematics Curriculum: The Scope of Personal Computational Technology. In A.J. Bishop, J. Clements, C. Keitel, J. Kilpatrick, and C. Laborde (Eds.), *International Handbook of Mathematics Education* (pp. 435–468). Dordrecht, the Netherlands: Kluwer.

Schifter, D., and Fosnot, C.T. (1993). *Reconstructing Mathematics Education: Stories of Teachers Meeting the Challenge of Reform.* New York: Teachers College Press.

Schmidt, W.H., McKnight, C.C., Cogan, L.S., Jakwerth, P.M., and Houang, R.T. (1999). *Facing the Consequences: Using TIMSS for a Closer Look at U.S. Mathematics and Science Education.* Dordrecht, The Netherlands: Kluwer Academic Publishers.

Schmidt, W.H., McKnight, C.C., Valverde, G.A., Houang, R.T., and Wiley, D.E. (1997). *Many Visions, Many Aims: A Cross-National Investigation of Curricular Intentions in School Mathematics.* Dordrecht, The Netherlands: Kluwer Academic Publishers.

Schoenfeld, A.H. (1985). *Mathematical Problem Solving.* Orlando, FL: Academic Press.

Smith, M. (2000). A Comparison of the Types of Mathematics Tasks and How They Were Completed During Eighth-Grade Mathematics Instruction in Germany, Japan, and the United States. Unpublished doctoral dissertation, University of Delaware.

Spindler, G.D. (Ed.), (1978). *The Making of Psychological Anthropology.* Berkeley: University of California Press.

Spindler, G.D., and Spindler, L. (1992). Cultural Process and Ethnography: An Anthropological Perspective. In M. LeCompte, W. Millroy, and J. Preisle (Eds.), *The Handbook of Qualitative Research in Education* (pp. 53–92). New York: Academic Press/Harcourt Brace.

Stanic, G.M.A., and Kilpatrick, J. (1988). Historical Perspectives on Problem Solving in the Mathematics Curriculum. In R.I. Charles and E.A. Silver (Eds.), *The Teaching and Assessing of Mathematical Problem Solving* (pp. 1–22). Reston, VA: National Council of Teachers of Mathematics.

Stein, M.K., Grover, B.W., and Henningsen, M. (1996). Building Student Capacity for Mathematical Thinking and Reasoning: An Analysis of Mathematical Tasks Used in Reform Classrooms. *American Educational Research Journal,* 33, 455–488.

Stein, M.K., and Lane, S. (1996). Instructional Tasks and the Development of Student Capacity to Think and Reason: An Analysis of the Relationship Between Teaching and Learning in a Reform Mathematics Project. *Educational Research and Evaluation,* 2 (1), 50–80.

Stigler, J.W., Gallimore, R., and Hiebert, J. (2000). Using Video Surveys to Compare Classrooms and Teaching Across Cultures: Examples and Lessons From the TIMSS Video Studies. *Educational Psychologist,* 35 (2), 87–100.

Stigler, J.W., Gonzales, P., Kawanaka, T., Knoll, S., and Serrano, A. (1999). *The TIMSS Videotape Classroom Study: Methods and Findings From an Exploratory Research Project on Eighth-Grade Mathematics Instruction in Germany, Japan, and the United States.* NCES 1999-074. Washington, DC: U.S. Department of Education, National Center for Education Statistics.

Stigler, J.W., and Hiebert, J. (1999). *The Teaching Gap: Best Ideas From the World's Teachers for Improving Education in the Classroom.* New York: Free Press.

Streefland, L. (1991). *Fractions in Realistic Mathematics Education: A Paradigm of Developmental Research.* Dordrecht, the Netherlands: Kluwer.

Tharp, R., and Gallimore, R. (1989). *Rousing Minds to Life: Teaching, Learning, and Schooling in Social Context.* Cambridge, England: Cambridge University Press.

UNESCO/OECD/EUROSTAT Data Collection. (2000). *2000 Data Collection on Education Statistics.*

Walberg, H. (1986). Synthesis of Research on Teaching. In M. Wittrock (Ed.), *Handbook of Research on Teaching* (3rd edition, pp. 214–229). New York: Macmillan.

Whitehead, A.N. (1948). *An Introduction to Mathematics.* New York: Oxford University Press.

Whiting, J.W.M. (1954). Methods and Problems in Cross-Cultural Research. In G. Lindzey (Ed.), *Handbook of Social Psychology* (pp. 523–531). Cambridge: Addison-Wesley.

Wittrock, M.C. (1986a). Students' Thought Processes. In M.C. Wittrock (Ed.), *Handbook of Research on Teaching* (3rd ed., pp. 297–314). New York: Macmillan.

Wittrock, M.C. (Ed.). (1986b). *Handbook of Research on Teaching* (3rd ed.). New York: Macmillan.

APPENDIX A
Sampling, Questionnaires, Video Data Coding Teams, and Statistical Analyses

A1. Sampling

The sampling objective for the TIMSS 1999 Video Study was to obtain a representative sample of eighth-grade mathematics lessons in each participating country.[1] Meeting this objective would enable inferences to be made about the national populations of lessons for the participating countries. In general, the sampling plan for the TIMSS 1999 Video Study followed the standards and procedures agreed to and implemented for the TIMSS 1999 assessments (see Martin, Gregory, and Stemler 2002). The school sample was required to be a Probability Proportionate to Size (PPS) sample. A PPS sample assigns probabilities of selection to each school proportional to the number of eligible students in the eighth-grade in schools countrywide. Then, one mathematics and/or one science eighth-grade class per school was sampled, depending on the subject(s) to be studied in each country.

Most of the participating countries drew separate samples for the Video Study and the assessments.[2] For this and other reasons, the TIMSS 1999 assessment data cannot be directly linked to the video database.[3]

A1.1. Sample Size

All of the TIMSS 1999 Video Study countries were required to include at least 100 schools in their initial selection of schools; however some countries chose to include more for various reasons. For example, Switzerland wished to analyze its data by language group, and therefore obtained a nationally representative sample that is also statistically reliable for the French-, Italian-, and German-language regions of that country. The Japanese data from the TIMSS 1995 Video Study included only 50 schools.

The TIMSS 1999 Video Study final sample included 638 eighth-grade mathematics lessons. Table 1 indicates the sample size and participation rate for each country.

---

[1] Australia, the Czech Republic, Japan, the Netherlands, and the United States also collected data on eighth-grade science lessons.
[2] For the German-speaking area of Switzerland, the video sample was a sub-sample of the TIMSS 1995 assessment schools. For Hong Kong SAR most, but not all, of the video sample was a sub-sample of the TIMSS 1999 assessment schools.
[3] Australia and Switzerland conducted separate studies that involved testing the mathematics achievement of the videotaped students.

Appendix A
Sampling, Questionnaires, Video Data Coding Teams, and Statistical Analyses

TABLE A.1. Sample size and participation rate for each country in the TIMSS 1999 Video Study

Country	Number of schools that participated	Percentage of schools that participated including replacements[1]— unweighted[2]	Percentage of schools that participated including replacements[1]— weighted[3]
Australia[4]	87	85	85
Czech Republic[4]	100	100	100
Hong Kong SAR	100	100	100
Japan[5]	50	100[6]	100[6]
Netherlands[4]	85[7]	87	85
Switzerland[8]	140	93	93
United States	83	77	76

[1]The participation rates including replacement schools are the percentage of all schools (i.e., original and replacements) that participated.
[2]Unweighted participation rates are computed using the actual numbers of schools and reflect the success of the operational aspects of the study (i.e., getting schools to participate).
[3]Weighted participation rates reflect the probability of being selected into the sample and describe the success of the study in terms of the population of schools to be represented.
[4]For Australia, the Czech Republic, and the Netherlands, these figures represent the participation rates for the combined mathematics and science samples.
[5]Japanese mathematics data were collected in 1995.
[6]The response rates after replacement for Japan differ from that reported previously (Stigler et al. 1999). This is because the procedure for calculating response rates after replacement has been revised to correspond with the method used in the TIMSS 1995 and TIMSS 1999 achievement studies.
[7]In the Netherlands, a mathematics lesson was filmed in 78 schools.
[8]In Switzerland, 74 schools participated from the German-language area (99 percent unweighted and weighted participation rate,), 39 schools participated from the French-language area (95 percent unweighted and weighted participation rate), and 27 schools participated from the Italian-language area (77 percent unweighted and weighted participation rate).
SOURCE: U.S. Department of Education, National Center for Education Statistics, Third International Mathematics and Science Study (TIMSS), Video Study, 1999.

A1.2. Sampling Within Each Country

Within the specified guidelines, the participating countries each developed their own strategy for obtaining a random sample of eighth-grade lessons to videotape for the TIMSS 1999 Video Study. For example, in two countries the video sample was a sub-sample of the TIMSS 1995 or TIMSS 1999 achievement study schools.[4]

The national research coordinators were responsible for selecting or reviewing the selection of schools and lessons in their country.[5] Identical instructions for sample selection were provided to all of the national research coordinators. For each country, a sample of at least 100 eighth-grade mathematics classrooms was selected for videotaping. National random samples of schools were drawn following the same procedure used to select the sample for the TIMSS 1999 main study. In all cases, countries provided the relevant sampling variables to Westat, so that they could appropriately weight the school samples.

Complete details about the sampling process in each country can be found in the technical report (Jacobs et al. forthcoming).

---

[4]For the German-language area of Switzerland, the video sample was a sub-sample of the TIMSS assessment schools. For Hong Kong SAR most, but not all, of the video sample was a sub-sample of the TIMSS 1999 assessment schools.
[5]Since it was based on the TIMSS 1999 assessment sample, the Hong Kong SAR school sample was selected and checked by Statistics Canada. In the United States, Westat selected the school sample and LessonLab selected the classroom sample.

A1.3. Videotaping Lessons

As noted in chapter 1, only one mathematics class was randomly selected within each school. No substitutions of teachers or class periods were allowed. The designated class was videotaped once, in its entirety, without regard to the particular mathematics topic being taught or type of activity taking place. The only exception was that teachers were not videotaped on days they planned to give a test for the entire class period.

Teachers were asked to do nothing special for the videotape session, and to conduct the class as they had planned. The scheduler and videographer in each country determined on which day the lesson would be filmed.

Most of the filming took place in 1999. In some countries filming began in 1998 and ended in 1999, and in others countries filming began in 1999 and ended in 2000. The goal was to sample lessons throughout a normal school year, while accommodating how academic years are organized in each country.

A2. Questionnaires

To help understand and interpret the videotaped lessons, questionnaires were collected from the eighth-grade mathematics teachers of each lesson. The teacher questionnaire was designed to elicit information about the professional background of the teacher, the nature of the mathematics course in which the lesson was filmed, the context and goal of the filmed lesson, and the teacher's perceptions of its typicality. Teacher questionnaire response rates are shown in table A.2.

TABLE A.2. Teacher questionnaire response rates

Country	Teacher questionnaire response rate (unweighted) Percent	Sample size
Australia	100	87
Czech Republic	100	100
Hong Kong SAR	100	100
Netherlands	96	75
Switzerland	99	138
United States	100	83

NOTE: Japan did not collect a new mathematics video sample for the TIMSS 1999 Video Study.
SOURCE: U.S. Department of Education, National Center for Education Statistics, Third International Mathematics and Science Study (TIMSS), Video Study, 1999.

The questionnaire was developed in English and consisted of 27 open-ended questions and 32 closed-ended questions. Each country could modify the questionnaire items to make them culturally appropriate. In some cases, questions were deleted from the questionnaires for reasons of sensitivity or appropriateness. Country-specific versions of the questionnaire were reviewed for comparability and accuracy. Additional details regarding the development of the questionnaire, along with a copy of the U.S. version of the teacher questionnaire, can be found in the technical report (Jacobs et al. forthcoming). Copies of the teacher questionnaires can also be found online at *http://www.lessonlab.com*.

The open-ended items in the teacher questionnaire required development of quantitative codes, a procedure for training coders, and a procedure for calculating inter-coder reliability. An 85 percent within-country inter-coder reliability criterion was used. The reliability procedures were similar to those used in the TIMSS 1995 assessment to code students' responses to the open-ended tasks (Mullis et al. 1996; Mullis and Martin 1998).

Short questionnaires also were distributed to the students in each videotaped lesson; however student data are not presented in this report. More information about the student questionnaire, and a copy of the U.S. version of the student questionnaire, can be found in the technical report (Jacobs et al. forthcoming).

A3. Video Data Coding

This section provides information about the teams involved in developing and applying codes to the video data. More details about each of these groups and the codes they developed and applied can be found in the technical report (Jacobs et al. forthcoming).

A3.1. The Mathematics Code Development Team

An international team was assembled to develop codes to apply to the TIMSS 1999 Video Study mathematics data. The team consisted of country associates (bilingual representatives from each country[6]) and was directed by a mathematics education researcher (see appendix B for team members). The mathematics code development team was responsible for creating and overseeing the coding process, and for managing the international video coding team. The team discussed coding ideas, created code definitions, wrote a coding manual, gathered examples and practice materials, designed a coder training program, trained coders and established reliability, organized quality control measures, consulted on difficult coding decisions, and managed the analyses and write-up of the data.

The mathematics code development team worked closely with two advisory groups: a group of national research coordinators representing each of the countries in the study, and a steering committee consisting of five North American mathematics education researchers (see appendix B for advisory group members).

A3.2. The International Video Coding Team

Members of the international video coding team represented all of the participating countries (see appendix B for team members). They were fluently bilingual so they could watch the lessons in their original language, and not rely heavily on the English-language transcripts. In almost all cases, coders were born and raised in the country whose lessons they coded.

---

[6]The mathematics team did not include a representative from Japan because Japanese mathematics lessons were not filmed as part of the TIMSS 1999 video data collection.

Coders in the international video coding team applied 45 codes in seven coding passes to each of the videotaped lessons. They also created a lesson table for each video, which combined information from a number of codes. For example, the lesson tables noted when each mathematical problem began and ended, and included a description of the problem and the solution. These tables served a number of purposes: they acted as quick reference guides to each lesson, they were used in the development process for later codes, and they enabled problems to be further coded by specialist coding teams.[7]

A3.3. Coding Reliability

As with any study that relies upon coding, it is important to establish clear reliability criteria. Based on procedures previously used and documented for the TIMSS 1995 Video Study and as described in the literature (Bakeman and Gottman 1997), percentage agreement was used to estimate inter-rater reliability and the reliability of codes within and across countries for all variables presented in the report. Percentage agreement allows for consideration of not only whether coders applied the same codes to a specific action or behavior, for example, but also allows for consideration of whether the coders applied the same codes within the same relative period of time during the lesson. That is, the reliability of coding in this study was judged based on two general factors: (1) that the same code was applied and (2) that it was applied during the same relative time segment in the lesson. Thus, it was not deemed appropriate to simply determine that the same codes were applied, but that they were applied to the same point in the lesson (here referred to as time segment) as well.

The calculation of percentage of agreement in this study is defined as the proportion of the number of agreements to the number of agreements and disagreements. Estimates of inter-rater and code reliability followed procedures described in Bakeman and Gottman (1997). Table A.3 reports the reliability of applying codes to the video data at two points: at or very near the beginning of applying codes (initial reliability) and at the midpoint of applying codes to the video data (midpoint reliability). Coders established initial reliability on all codes in a coding pass prior to their implementation. After the coders finished coding approximately half of their assigned set of lessons (in most cases about 40 to 50 lessons), coders established midpoint reliability. The minimum acceptable reliability score for each code (averaging across coders) was 85 percent. Individual coders or coder pairs had to reach at least 80 percent reliability on each code.[8]

Initial reliability was computed as agreement between coders and a master document. A master document refers to a lesson or part of a lesson coded by consensus by the mathematics code development team. To create a master, the country associates independently coded the same lesson and then met to compare their coding and discuss disagreements until consensus was achieved. Masters were used to establish initial reliability. This method is considered a rigorous and cost-effective alternative to inter-coder reliability (Bakeman and Gottman 1997).

---

[7] A subset of these lesson tables, from all countries except Japan, were expanded and then coded by the mathematics quality analysis group, described below.

[8] The minimum acceptable reliability score for all codes (across coders and countries) was 85 percent. For coders and countries, the minimum acceptable reliability score was 80 percent. That is, the reliability of an individual coder or the average of all coders within a particular country was occasionally between 80 and 85 percent. In these cases clarification was provided as necessary, but re-testing for reliability was not deemed appropriate.

Midpoint reliability was computed as agreement between pairs of coders. By halfway through the coding process, coders were considered to be more expert in the code definitions and applications than the mathematics code development team. Therefore, in general, the most appropriate assessment of their reliability was deemed to be a comparison among coders rather than to a master document. Inter-rater agreement was also used to establish initial reliability in some of the later coding passes, but only for those codes for which coders helped to develop coding definitions.

A percentage agreement reliability statistic was computed for each coder by dividing the number of agreements by the sum of agreements and disagreements (Bakeman and Gottman 1997). Average reliability was then calculated across coders and across countries for each code. In cases where coders did not reach the established reliability standard, they were re-trained and re-tested using a new set of lessons. Codes were dropped from the study if 85 percent reliability could not be achieved (or if individual coders could not reach at least 80 percent reliability). As indicated in table A.3, all codes presented in the report met or exceeded the minimum acceptable reliability standard established for this study.

What counted as an agreement or disagreement depended on the specific nature of each code, and is explained in detail in Jacobs et al. (forthcoming). Some codes required coders to indicate a time. In these cases, coders' time markings had to fall within a predetermined margin of error. This margin of error varied depending on the nature of the code, ranging from 10 seconds to 2 minutes. Rationales for each code's margin of error are provided in Jacobs et al. (forthcoming).

Exact agreement was required for codes that had categorical coding options. In other words, if a code had four possible coding categories, coders had to select the same coding category as the master. In most cases, coders had to both mark a time (i.e., note the in- and/or out-point of a particular event) and designate a coding category. In these cases, it was first determined whether coders reliably marked the same or nearly the same in- and out-points, within the established margin of error. If reliability could not be established between coders based on marking the in- and out-time of codes, then reliability for the actual coding category was not calculated. In these cases, as explained above, coders were re-trained and re-tested using a different set of lessons.

Percentage agreement was used to estimate inter-rater reliability and the reliability of the codes within and across countries for all the variables presented in this report. Percentage agreement allowed us to take into account the markings of both in- and out-points of the codes applied to the videotaped lessons when computing the reliability for a code. All three marks (i.e., in-point, out-point, and label) were included in the calculation. Percentage agreement was selected to calculate reliability for all codes because most codes included marking times as well as labels.

While initial and midpoint reliability rates are reported, coders were monitored throughout the coding process to avoid reliability decay. If a coder did not meet the minimum reliability standard, additional training was provided until acceptable reliability was achieved. The data reported in the report only include data from coders who were evaluated as reliable.

Table A.3 lists the initial and midpoint reliability scores for each code, averaged across coders.

TABLE A.3. Initial and midpoint reliability statistics for each code applied by the International Coding Team, by code: 1999

Code	Initial reliability[1] (percent)	Midpoint reliability[2] (percent)
Lesson (LES)	93	99
Classroom interaction (CI)	94	92
Content activity (CC)	90	87
Concurrent problem (CP)	94	90
Assignment of homework (AH)	99	93
Goal statement (GS)	99	89
Outside interruption (OI)	96	96
Summary of lesson (SL)	98	99
Homework (H)	99	98
Real-life connection (RLC)	98	100
Graphs (GR)	97	98
Tables (TA)	99	98
Drawings/diagrams (DD)	97	94
Physical materials (PM)	95	97
Student choice of solution method (SC)	90	93
Proof/verification/derivation (PVD)	99	97
Number of target results (NTR)	96	94
Length of working on (LWO)	95	94
Facilitating exploration (FE)	96	95
Chalkboard (CH)	96	100
Projector (PRO)	98	100
Television or video (TV)	100	100
Textbook or worksheets (TXW)	98	98
Special mathematical materials (SMM)	92	93
Real-world objects (RWO)	98	100
Calculators (CALC)	98	95
Computers (COMP)	100	98
Multiple solution methods (MSM)	99	98
Problem summary (PSM)	97	95
Contextual information (CON)	92	91
Mathematical concept/theory/idea (CTI)	92	94
Activity (AC)	97	97
Announcing or clarifying homework or test (HT)	95	98
Private work assignment (PWA)	93	98
Organization of students (OS)	96	96
Public announcements (PA)	86	86
Purpose (P)	87	94

[1] Initial reliability refers to reliability established on a designated set of lessons before coders began work on their assigned lessons.
[2] Midpoint reliability refers to reliability established on a designated set of lessons after coders completed approximately half of their total assigned lessons.
SOURCE: U.S. Department of Education, National Center for Education Statistics, Third International Mathematics and Science Study (TIMSS), Video Study, 1999.

A variety of additional quality control measures were put in place to ensure accurate coding. These measures included: (1) discussing difficulties in coding reliability lessons with the mathematics code development team and/or other coders, (2) checking the first two lessons coded by each coder, either by a code developer or by another coder, and (3) discussing hard-to-code lessons with code developers and/or other coders.

A3.4. Specialist Coding Groups

The majority of codes for which analyses were conducted for in this report were applied to the video data by members of the international video coding team, who were cultural "insiders" and fluent in the language of the lessons they coded. However, not all of them were experts in mathematics or teaching. Therefore, several specialist coding teams with different areas of expertise were employed to create and apply special codes regarding the mathematical nature of the content, the pedagogy, and the discourse.

A3.4.1. Mathematics Problem Analysis Group

The mathematics problem analysis group was comprised of individuals with expertise in mathematics and mathematics education (see appendix B for group members). They developed and applied a series of codes to all of the mathematical problems in the videotaped lessons, using lesson tables prepared by the international video coding team.

The mathematics problem analysis group constructed a comprehensive, detailed, and structured list of mathematical topics covered in eighth grade in all participating countries. Each problem marked in a lesson was connected to a topic on the list.

In addition to coding the mathematical topic of problems, the group also coded the procedural complexity of each problem, the relationship among problems, and identified application problems (see chapter 4 for definitions of procedural complexity and problem relationship, and chapter 5 for the definition of application problems).

The members of this group each established reliability with the director of the group by coding a randomly selected set of lessons from each country. They computed initial reliability as well as reliability after approximately two-thirds of the lessons had been coded. The percent agreement was above 85 percent for each code at both time points.

The director prepared a "master" for each lesson. Table A.4 lists the other coders' percentage agreement with the director on each code, calculated as the number of agreements divided by the sum of agreements and disagreements.

TABLE A.4. Initial and midpoint reliability statistics for each code applied by the Mathematics Problem Analysis Group, by code: 1999

Code	Initial reliability (percent)[1]	Midpoint reliability (percent)[2]
Topic	89	90
Procedural complexity	87	90
Relationship	88	88

[1]Initial reliability refers to reliability established on a designated set of lessons before coders began work on their assigned lessons.
[2]Midpoint reliability refers to reliability established on a designated set of lessons after coders completed approximately two-thirds of their assigned lessons.
SOURCE: U.S. Department of Education, National Center for Education Statistics, Third International Mathematics and Science Study (TIMSS), Video Study, 1999.

A3.4.2. Mathematics Quality Analysis Group

A second specialist group possessed special expertise in mathematics and teaching mathematics at the post-secondary level (see appendix B for group members). The same group previously was commissioned to develop and apply codes for the TIMSS 1995 Video Study. The mathematics quality analysis group reviewed a randomly selected subset of 120 lessons (20 lessons from each country except Japan). Japan was not included in this exercise because the group already had analyzed a sub-sample of the Japanese lessons as part of the 1995 Video Study.

Specially trained members of the international video coding team created expanded lesson tables for each lesson in this subset. The resulting 120 tables all followed the same format: they included details about the classroom interaction, the nature of the mathematical problems worked on during class time, descriptions of time periods during which problems were not worked on, mathematical generalizations, labels, links, goal statements, lesson summaries, and other information relevant to understanding the content covered during the lesson. Furthermore, the tables were "country-blind," with all indicators that might reveal the country removed. For example, "pesos" and "centavos" were substituted as units of currency, proper names were changed to those deemed neutral to Americans, and lessons were identified only by an arbitrarily assigned ID number. The mathematics quality analysis group worked solely from these written records, and had no access to either the full transcript or the video data.

The mathematics quality analysis group created and applied a coding scheme that focused on mathematical reasoning, mathematical coherence, the nature and level of mathematical content, and the overall quality of the mathematics in the lessons. The scheme was reviewed by mathematics experts in each country and then revised based on the feedback received. The group applied their coding scheme by studying the written records of the lessons and reaching consensus about each judgment. Due to the small sample size, only descriptive analyses of the group's coded data are included in this report (see appendix D).

A3.4.3. Problem Implementation Analysis Team

The problem implementation analysis team analyzed a subset of mathematical problems and examined (1) the types of mathematical processes implied by the problem statement and (2) the types of mathematical processes that were publicly addressed when solving the problem (see appendix B for group members).

Using the video data, translated transcripts, and the same lesson tables provided to the mathematics problem analysis group, the problem implementation analysis team analyzed only those problems that were publicly completed during the videotaped lesson. Problems had to be publicly completed in order for the group to code for problem implementation. Furthermore, the group did not analyze data from Switzerland, since most of the Swiss transcripts were not translated into English.

Reliability was established by comparing a set of 10 lessons from each country coded by the director of the team with one outside coder. These lessons were randomly selected from those lessons that included at least one problem that was publicly completed during the lesson. Reliability of at least 85 percent was achieved for all countries.

Average inter-rater agreement for problem statements and implementations is shown in table A.5. Percentage agreement was calculated as the number of agreements divided by the sum of agreements and disagreements.

TABLE A.5. Reliability statistics for each code applied by the Problem Implementation Analysis Group, by code: 1999

Code	Reliability (percent)
Problem statement	90
Problem implementation	90

SOURCE: U.S. Department of Education, National Center for Education Statistics, Third International Mathematics and Science Study (TIMSS), Video Study, 1999.

A3.4.4. Text Analysis Group

The text analysis group used all portions of the mathematics lesson transcripts designated as public interaction to conduct various discourse analyses (see appendix B for group members). The group utilized specially designed computer software for these quantitative analyses of classroom talk.

Because of resource limitations, computer-assisted analyses were applied to English translations of lesson transcripts.[9] In the case of the Czech Republic, Japan, and the Netherlands all lessons were translated from the respective native languages, and in the case of Hong Kong SAR, 34 percent of the lessons were conducted in English, so 66 percent were translated. English translations of Swiss lessons were not available.

---

[9] Transcriber/translators were fluent in both English and their native language, educated at least through eighth grade in the country whose lessons they translated, and had completed two-weeks training in the procedures detailed in the TIMSS 1999 Video Study Transcription and Translation Manual (available in Jacobs et al. forthcoming). A glossary of terms was developed to help standardize translation within each country.

A4. Statistical Analyses

Most of the analyses presented in this report are comparisons of means or distributions across seven countries for video data and across six countries for questionnaire data. The TIMSS 1999 Video Study was designed to provide information about and compare mathematics instruction in eighth-grade classrooms. For this reason, the lesson rather than the school, teacher, or student was the unit of analysis in all cases.

Analyses were conducted in two stages. First, means or distributions were compared across all available countries using either one-way ANOVA or Pearson Chi-square procedures. For some continuous data, additional dichotomous variables were created that identified either no occurrence of an event (code = 0) or one or more occurrences of an event (code = 1). Variables coded dichotomously were usually analyzed using ANOVA, with asymptotic approximations.

Next, for each analysis that was significant overall, pairwise comparisons were computed and significance determined by the Bonferroni adjustment. The Bonferroni adjustment was made assuming all combinations of pairwise comparisons. For continuous variables, Student's t values were computed on each pairwise contrast. Student's t was computed as the difference between the two sample means divided by the standard error of the difference. Determination that a pairwise contrast was statistically significant with $p<.05$ was made by consulting the Bonferroni t tables published by Bailey (1977). For categorical variables, the Bonferroni Chi-square tables published in Bailey (1977) were used.

The degrees of freedom were based on the number of replicate weights, which was 50 for each country. Thus, in any comparison between two countries there were 100 replicate weights, which were used as the degrees of freedom.

A significance level criterion of .05 was used for all analyses. All differences discussed in this report met at least this level of significance, unless otherwise stated. Terms such as "less," "more," "greater," "higher," or "lower," for example, are applied only to statistically significant comparisons. The inability to find statistical significance is noted as "no differences detected." In some cases, large apparent differences in data are not significant due to large standard errors, small sample sizes, or both.

All tests were two-tailed. Statistical tests were conducted using unrounded estimates and standard errors, which also were computed for each estimate. Standard errors for estimates shown in figures in the report are provided in appendix C.

The analyses reported here were conducted using data weighted with survey weights, which were calculated specifically for the classrooms in the TIMSS 1999 Video Study (see Jacobs et al. forthcoming for a more detailed description of weighting procedures).

APPENDIX B
Participants in the TIMSS 1999 Video Study of Mathematics Teaching

Director of TIMSS 1999 Video Study of Mathematics Teaching

James Hiebert

Directors of TIMSS 1999 Video Study

Ronald Gallimore
James Stigler

National Research Coordinators

Australia
  Jan Lokan
  Barry McCrae
Czech Republic
  Jana Strakova
Hong Kong SAR
  Frederick Leung
Japan
  Shizuo Matsubara
  Yasushi Ogura
  Hanako Senuma
Netherlands
  Klaas Bos
Switzerland
  Primary: Kurt Reusser
  Swiss-French: Norberto Bottani
  Swiss-German: Christine Pauli
  Swiss-Italian: Emanuele Berger
United States
  Patrick Gonzales

U.S. Steering Committee

Thomas Cooney
Douglas Grouws
Carolyn Kieran
Glenda Lappan
Edward Silver

Chief Analyst

Helen Garnier

Country Associates/Mathematics Code Development Team

Australia
  Hilary Hollingsworth
Czech Republic
  Svetlana Trubacova
Hong Kong SAR
  Angel Chui
  Ellen Tseng
Netherlands
  Karen Givvin
Switzerland
  Nicole Kersting
United States
  Jennifer Jacobs

Teams Working Outside the United States

Australia
  Brian Doig
  Silvia McCormack
  Gayl O'Connor
Switzerland
  Kathya Tamagni Bernasconi
  Giulliana Cossi
  Olivier de Marcellus
  Ruhal Floris
  Isabelle Hugener
  Narain Jagasia
  Miriam Leuchter
  Francesca Pedrazzini-Pesce
  Dominik Petko
  Monika Waldis

Questionnaire Coding Team

Helen Garnier
Margaret Smith

Mathematics Quality Analysis Team

Phillip Emig
Wallace Etterbeek
Alfred Manaster
Barbara Wells

Appendix B
Participants in the TIMSS 1999 Video Study of Mathematics Teaching

Problem Implementation Analysis Team

Christopher S. Hlas
Margaret Smith

Mathematics Problem Analysis Team

Eric Sisofo
Margaret Smith
Diana Wearne

Field Test Team

Karen Givvin
Jennifer Jacobs
Takako Kawanaka
Christine Pauli
Jean-Paul Reeff
Nick Scott
Svetlana Trubacova

Questionnaire Development Team

Sister Angelo Collins
Helen Garnier
Kass Hogan
Jennifer Jacobs
Kathleen Roth

Text Analysis Group

Steve Druker
Don Favareau
Takako Kawanaka
Bruce Lambert
David Lewis
Fang Liu
Samer Mansukhani
Genevieve Patthey-Chavez
Rodica Waivio
Clement Yu

Chief Videographers

Maria Alidio
Takako Kawanaka
Scott Rankin

Videographers

Sue Bartholet
Talegon Bartholet
Michaela Bractalova
Gabriel Charmillot
Matthias Feller
Ruud Gort
Christopher Hawkins
Kurt Hess
Rowan Humphrey
Narian Jagasia
LeAnne Kline
Tadayuki Miyashiro
Silvio Moro
Selin Ondül
Mike Petterson
Stephen Skok
Sikay Tang
Giovanni Varini
Sofia Yam
Jiri Zeiner
Andreas Zollinger

Field Test Videographer

Ron Kelly

Computer Programming

Paul Grudnitski
Daniel Martinez
Ken Mendoza
Carl Manaster
Rod Kent

Consultants

Justus J. Schlichting
Genevieve Patthey-Chavez
Takako Kawanaka
Rossella Santagata

Transcription and Translation Directors

Lindsey Engle
Don Favareau
Wendy Klein

Petra Kohler
David Olsher
Susan Reese

Transcribers and Translators

Australia
   Marco Duranti
   Hugh Grinstead
   Amy Harkin
   Tammy Lam
   Tream Anh Le Duc
   James Monk
   Aja Stanman
   Elizabeth Tully
   Daniella Wegman

Czech Republic
   Barbara Brown
   Jana Hatch
   Peter Kasl
   Vladimir Kasl
   Jirina Kyas
   Alena Mojhova
   Vaclav Plisek

Hong Kong SAR
   Stella Chow
   Khue Duong
   Ka (Fiona) Ng
   Constance Wong
   Jacqueline Woo
   Kin Woo
   Lilia Woo

Japan
   Kaoru Koda
   Yuri Kusuyama
   Ken Kuwabara
   Emi Morita
   Angela Nonaka
   Naoko Otani
   Jun Yanagimachi

Netherlands
   Hans Angessens
   Tony DeLeeuw
   Neil Galanter
   Maaike Jacobson
   Maarten Lobker
   Yasmin Penninger
   Linda Pollack

   Silvia Van Dam

Switzerland
   Michele Balmelli
   Valérie Bourdet
   Laura Cannon
   Chiara Ceccarelli
   Ankara Chen
   Sabina del Grosso
   Anne de Preux
   Andrea Erzinger
   Maria Ferraiuolo
   Véronique Gendre
   Ueli Halbheer
   Sandra Hay
   Nicole Kersting
   Petra Kohler
   Leata Kollart
   Nur Kussan
   Anita Lovasz
   Malaika Mani
   Céline Marco
   Annamaria Mauramatti
   Giovanni Murialdo
   Luisa Murialdo
   Philippe Poirson
   Astrid Sigrist
   Doris Steinemann
   Simona Torriani
   Emilie Tran
   Mélanie Tudisco
   Nathalie Vital
   Esther Wertmüller
   Rebekka Wyss
   Salome Zahn

United States
   Ginger (Yen) Dang
   Jake Elsas
   Jordan Engle
   Steven Gomberg
   Barry Griner
   Jaime Gutierrez
   Sydjea Johnson
   Keith Murphy
   Kimberly Nelson
   Raoul Rolfes
   Tosha Schore
   Budie Suriawidjaja

Administrative Support

Maria Alidio
Cori Busch
Ellen Chow
Olivier de Marcellus
Melanie Fan
Tammy Haber
Christina Hartmann
Gail Hood
Rachael Hu
Brenda Krauss
Samuel Lau
Vicki Momary
Francesca Pedrazzini-Pesce
Liz Rosales
Rossella Santagata
Eva Schaffner
Yen-lin Schweitzer
Cynthia Simington
Kathya Tamagni-Bernasconi
Vik Thadani
Jennifer Thomas-Hollenbeck
Sophia Yam
Andreas Zollinger

Video Processing

Don Favareau
Tammy Haber
Petra Kohler
Brenda Krauss
Miriam Leuchter
David Martin
Alpesh Patel
Susan Reese
Liz Rosales
Steven Schweitzer

School Recruiters

Australia
   Silvia McCormack
United States
   Marty Gale
   Elizabeth Tully

International Video Coding Team

Australia
   David Rasmussen
   Amanda Saw
Czech Republic
   Jana Hatch
   Alena Mojhova
   Tom Hrouda
Hong Kong SAR
   Angel Chui
   Ellen Tseng
Japan
   Jun Yanagimachi
   Kiyomi Chinen
Netherlands
   Yasmin Penninger
   Linda Pollack
   Dagmar Warvi
Switzerland
   Giuliana Cossi
   Natascha Eckstein
   Domenica Flütsch
   Isabelle Hugener
   Kathrin Krammer
   Anita Lovasz
   Rossella Santagata
   Ursula Schwarb
   Regina Suhner
   Monika Waldis
United States
   Samira Rastegar
   Bayard Lyons
   Girlie Delacruz
   Budie Suriawidjaja

Mathematics Education Experts

Australia
   Judy Anderson
   Rosemary Callingham
   David Clarke
   Neville Grace
   Alistair McIntosh
   Will Morony
   Nick Scott
   Max Stephens

Czech Republic
  Jiri Kadlecek
  Daniel Pribik
  Vlada Tomasek
  Miloslav Fryzek
  Iveta Kramplova
Hong Kong SAR
  P.H. Cheung
  C.I. Fung
  Fran Lopez-Real
  Ida Mok
  M.K. Siu
  N.Y. Wong
Japan
  Hanako Senuma
  Kazuhiko Souma
Switzerland
  Gianfranco Arrigo
  Susan Baud
  Pierre Burgermeister
  Alain Correvon
  Olivier de Marcellus
  Kurt Eggenberger
  Ruhal Floris
  Christian Rorbach
  Beat Wälti
United States
  Thomas Cooney
  Megan Franke
  Douglas Grouws
  Carolyn Kieran
  Magdalene Lampert
  Glenda Lappan
  Eugene Owen
  Edward Silver
  Diana Wearne

Typical Lesson Analysis

Australia
  Barry McCrae
  Alistair McIntosh
  Will Morony
Hong Kong SAR
  P.H. Cheung
  T.W. Fung
  K.K. Kwok
  Arthur Lee
  Ida Mok
  K.L. Wong
  N.Y. Wong
  Patrick Wong
United States
  Thomas Cooney
  Douglas Grouws
  Carolyn Kieran
  Glenda Lappan
  Edward Silver

Westat (Weights and Sampling)

Justin Fisher
Susan Fuss
Mary Nixon
Keith Rust
Barbara Smith-Brady
Ngoan Vo

Design and Layout

Fahrenheit Studio

APPENDIX C
Standard Errors for Estimates Shown in Figures and Tables

Standard errors for estimates shown in figures and tables, by country[1]

Table/Figure	Category	AU	CZ	HK	JP[2]	NL	SW	US
Table 2.1	Mathematics	0.1	0.0	0.1	—	0.0	0.1	0.1
	Science	0.0	0.1	0.1	—	0.1	0.1	0.1
	Education	0.1	0.0	0.0	—	0.0	0.0	0.1
	Other	0.1	0.0	0.1	—	0.1	0.0	0.1
Table 2.2	Certified to teach one or more areas	0.0	0.0	0.0	—	0.0	0.0	0.0
	Certified to teach mathematics at grade 8	0.1	0.0	0.0	—	0.0	0.0	0.1
	Certified to teach mathematics at another or unspecified grade	0.0	0.0	0.0	—	0.0	0.0	0.0
	Certified to teach science at grade 8	0.1	0.0	0.1	—	0.1	0.0	0.1
	Certified to teach science at another or unspecified grade	0.0	0.0	0.0	—	0.0	0.0	0.1
	Certified to teach education at grade 8	0.0	0.0	0.0	—	0.0	0.0	0.1
	Certified to teach education at another or unspecified grade	0.0	0.0	0.0	—	0.0	0.0	0.0
	Certified to teach another subject at grade 8	0.0	0.0	0.1	—	0.0	0.0	0.0
	Certified to teach another subject at another or unspecified grade	0.0	0.0	0.0	—	0.0	0.0	0.0
Table 2.3	Years teaching	1.1	1.1	0.8	—	1.1	1.0	1.2
	Years teaching mathematics	1.1	1.2	0.8	—	1.0	1.1	1.3
Table 2.4	Hours per week teaching mathematics	0.7	0.5	0.5	—	0.8	0.4	1.2
	Hours per week teaching other classes	0.7	0.5	0.6	—	0.7	0.5	0.9
	Hours per week meeting with other teachers	0.2	0.2	0.1	—	0.2	0.1	0.3
	Hours per week doing mathematics-related work at school	0.7	0.4	0.6	—	0.4	0.2	0.4
	Hours per week doing mathematics-related work at home	0.6	0.3	0.5	—	0.5	0.3	0.5
	Hours per week doing other school-related activities	0.8	1.0	0.8	—	0.6	0.7	0.7
	Hours per week teaching and doing other school-related activities	1.1	1.2	1.4	—	1.3	0.7	1.4
Figure 2.1	Perspective goals	0.1	0.0	0.0	—	0.0	0.0	0.0
	Process goals	0.1	0.0	0.0	—	0.0	0.0	0.0
	Content goals	0.1	0.0	0.0	—	0.0	0.0	0.0

Appendix C
Standard Errors for Estimates Shown in Figures and Tables

Standard errors for estimates shown in figures and tables, by country[1]—Continued

Table/Figure	Category	AU	CZ	HK	JP2	NL	SW	US
Table 2.5	Using routine operations	0.1	0.1	0.1	—	0.1	0.0	0.1
	Reasoning mathematically	0.0	0.0	0.0	—	0.1	0.0	0.0
	Applying mathematics to real-world problems	0.0	0.0	0.0	—	0.0	0.0	0.1
	Knowing mathematical content	0.1	0.0	0.0	—	0.0	0.0	0.0
	No process goals identified	0.1	0.0	0.0	—	0.0	0.0	0.0
Table 2.6	Curriculum guidelines	0.0	0.0	0.1	—	0.1	0.0	0.1
	External exams or tests	—	0.0	0.1	—	0.0	0.1	0.1
	Mandated textbook	0.1	0.0	0.1	—	0.0	0.0	0.1
	Teacher's comfort with or interest in the topic	0.1	0.0	0.1	—	0.0	0.1	0.1
	Teacher's assessment of students' interests or needs	0.1	0.1	0.1	—	0.0	0.0	0.1
	Cooperative work with other teachers	0.1	0.0	0.0	—	0.1	0.0	0.0
Figure 2.2	Disagree	0.0	0.0	0.0	—	0.0	0.0	0.0
	No opinion	0.1	0.0	0.0	—	0.0	0.0	0.0
	Agree	0.1	0.0	0.0	—	0.0	0.1	0.1
Figure 2.3	Not at all	0.0	0.0	0.1	—	0.0	0.0	0.0
	A little	0.1	0.1	0.0	—	0.1	0.0	0.0
	A fair amount or a lot	0.1	0.1	0.0	—	0.1	0.0	0.0
Figure 2.4	Sometimes or seldom	0.1	0.0	0.0	—	0.1	0.0	0.1
	Often	0.1	0.0	0.1	—	0.1	0.1	0.1
	Almost always	0.1	0.0	0.0	—	0.1	0.0	0.1
Figure 2.5	Worse than usual	0.0	0.1	0.0	—	0.1	0.0	0.0
	About the same	0.1	0.1	0.1	—	0.1	0.0	0.1
	Better than usual	0.1	0.0	0.1	—	0.0	0.0	0.1
Figure 2.6	Less difficult	0.0	0.0	0.0	—	0.0	0.0	0.1
	About the same	0.0	0.0	0.0	—	0.1	0.1	0.1
	More difficult	0.0	0.0	0.0	—	0.0	0.0	0.0
Figure 2.7	Worse than usual	0.0	0.1	0.1	—	0.0	0.0	0.0
	About the same	0.0	0.0	0.0	—	0.0	0.0	0.0
	Better than usual	0.0	0.0	0.0	—	0.0	0.0	0.0
Figure 2.8	Time spent planning similar lessons	2.1	1.6	1.7	—	0.8	2.5	3.1
	Time spent planning the videotaped lesson	5.1	3.9	7.0	—	2.1	2.3	5.4

Standard errors for estimates shown in figures and tables, by country[1]—Continued

Table/Figure	Category	AU	CZ	HK	JP[2]	NL	SW	US
Table 2.7	Number of lessons in the unit	0.6	0.8	0.5	—	0.9	1.1	0.6
	Placement in the unit	0.5	0.7	0.4	—	0.6	0.9	0.5
Table 3.1	Lesson duration – average	1.4	0.1	1.4	0.3	0.8	0.4	1.9
	Lesson duration – standard deviation	1.0	0.2	1.5	0.3	2.0	3.1	2.3
Figure 3.2	Mathematical work	0.0	0.0	0.0	0.0	0.0	0.0	0.0
	Mathematical organization	0.0	0.0	0.0	0.0	0.0	0.0	0.0
	Non-mathematical work	0.0	0.0	0.0	0.0	0.0	0.0	0.0
Figure 3.3	Problem segments	0.0	0.0	0.0	0.0	0.0	0.0	0.0
	Non-problem segments	0.0	0.0	0.0	0.0	0.0	0.0	0.0
Table 3.3	Independent problems – number	1.9	1.0	0.5	0.1	1.1	0.5	1.0
	Answered-only problems – number	1.0	0.1	0.1	0.2	0.6	1.5	1.4
Figure 3.4	Independent problems – percent of lesson time	0.0	0.0	0.0	0.1	0.0	0.0	0.0
	Concurrent problems – percent of lesson time	0.0	0.0	0.0	0.1	0.0	0.0	0.0
	Answered-only problems – percent of lesson time	0.0	0.0	0.0	0.0	0.0	0.0	0.0
Figure 3.5	Time per independent problem	0.5	0.3	0.4	1.5	0.3	0.6	1.1
Figure 3.7	Independent and concurrent problems longer than 45 seconds	0.0	0.0	0.0	0.0	0.0	0.0	0.0
Figure 3.8	Reviewing – percent of lesson time	0.0	0.0	0.0	0.0	0.1	0.0	0.0
	Introducing new content – percent of lesson time	0.0	0.0	0.0	0.0	0.0	0.0	0.0
	Practicing new content – percent of lesson time	0.0	0.0	0.0	0.0	0.0	0.0	0.0
Table 3.4	Reviewing – at least one segment	0.0	0.0	0.0	0.1	0.1	0.0	0.0
	Introducing new content – at least one segment	0.1	0.0	0.0	0.0	0.0	0.0	0.1
	Practicing new content – at least one segment	0.1	0.0	0.0	0.1	0.1	0.0	0.1
Figure 3.9	Entirely review	0.1	0.0	0.0	0.0	0.1	0.0	0.1
Table 3.5	Shifts in purpose	0.2	0.1	0.2	0.1	0.1	0.1	0.2
Table 3.6	Public interaction	0.0	0.0	0.0	0.0	0.0	0.0	0.0
	Private interaction	0.0	0.0	0.0	0.0	0.0	0.0	0.0
	Optional, student presents information	0.0	0.0	0.0	0.0	0.0	0.0	0.0
Figure 3.10	Worked individually	0.1	0.0	0.0	0.0	0.0	0.0	0.0
	Worked in pairs and groups	0.1	0.0	0.0	0.0	0.0	0.0	0.0

Appendix C
Standard Errors for Estimates Shown in Figures and Tables

Standard errors for estimates shown in figures and tables, by country[1]—Continued

Table/Figure	Category	AU	CZ	HK	JP[2]	NL	SW	US
Table 3.7	Shifts in classroom interaction	0.3	0.4	0.3	0.4	0.3	0.3	0.4
Figure 3.11	Homework assigned	0.1	0.0	0.0	0.1	0.1	0.0	0.1
Table 3.8	Future homework problems – number	1.4	0.2	0.5	0.2	1.5	0.6	1.0
	Future homework problems – time	0.9	0.4	0.7	0.6	1.4	1.2	0.8
Table 3.9	Previous homework problems – number	1.7	0.1	0.1	0.2	1.8	1.9	1.9
	Previous homework problems – time	0.4	0.1	0.3	0.4	3.4	1.8	2.6
Figure 3.12	Goal statement	0.0	0.0	0.0	0.1	0.1	0.1	0.1
Figure 3.13	Summary statement	0.0	0.0	0.0	0.1	0.0	0.0	0.0
Figure 3.14	Outside interruption	0.1	0.0	0.0	0.1	0.1	0.0	0.1
Figure 3.15	Non-mathematical segment within the mathematics portion of the lesson	0.0	0.0	0.0	0.0	0.1	0.0	0.0
Figure 3.16	Public announcement unrelated to the current assignment	0.0	0.0	0.0	0.1	0.1	0.0	0.0
Table 4.1	Number	0.1	0.0	0.0	0.0	0.0	0.0	0.0
	Whole numbers, fractions, decimals	0.1	0.0	0.0	0.0	0.0	0.0	0.0
	Ratio, proportion, percent	0.0	0.0	0.0	0.0	0.0	0.0	0.0
	Integers	0.0	0.0	0.0	0.0	0.1	0.0	0.1
	Geometry	0.1	0.0	0.0	0.0	0.1	0.0	0.0
	Measurement	0.0	0.0	0.0	0.1	0.0	0.0	0.0
	Two-dimensional geometry	0.0	0.0	0.0	0.1	0.0	0.0	0.0
	Three-dimensional geometry	0.0	0.0	0.0	0.1	0.0	0.0	0.0
	Statistics	0.0	0.0	0.0	0.0	0.0	0.0	0.0
	Algebra	0.0	0.0	0.0	0.0	0.1	0.0	0.1
	Linear expressions	0.0	0.0	0.0	0.0	0.0	0.0	0.0
	Solutions and graphs of linear equations and inequalities	0.0	0.0	0.0	0.0	0.1	0.0	0.1
	Higher-order functions	0.0	0.0	0.0	0.0	0.0	0.0	0.0
	Trigonometry	0.0	0.0	0.0	0.0	0.0	0.0	0.0
	Other	0.0	0.0	0.0	0.0	0.0	0.0	0.0
Figure 4.1	Low complexity problems	0.0	0.0	0.0	0.0	0.0	0.0	0.1
	Moderate complexity problems	0.0	0.0	0.0	0.1	0.0	0.0	0.0
	High complexity problems	0.0	0.0	0.0	0.1	0.0	0.0	0.0

Standard errors for estimates shown in figures and tables, by country[1]—Continued

Table/Figure	Category	AU	CZ	HK	JP[2]	NL	SW	US
Figure 4.2	Low complexity 2-D geometry problems	0.1	0.1	0.1	0.0	0.1	0.1	0.2
	Moderate complexity 2-D geometry problems	0.1	0.1	0.1	0.0	0.1	0.1	0.2
	High complexity 2-D geometry problems	0.1	0.1	0.0	0.1	0.1	0.0	0.1
Figure 4.3	Proof – percent of problems	0.0	0.0	0.0	0.1	0.0	0.0	0.0
Figure 4.4	Proof – at least one	0.0	0.0	0.0	0.1	0.0	0.0	0.0
Figure 4.5	Proof – percent of 2-D geometry problems	0.0	0.0	0.0	0.1	0.0	0.0	0.0
Figure 4.6	Unrelated problems	0.0	0.0	0.0	0.0	0.0	0.0	0.0
	Repetition problems	0.0	0.0	0.0	0.0	0.0	0.0	0.0
	Thematically related problems	0.0	0.0	0.0	0.0	0.0	0.0	0.0
	Mathematically related problems	0.0	0.0	0.0	0.1	0.0	0.0	0.0
Table 4.2	Unrelated – number of problems	0.1	0.1	0.1	0.0	0.1	0.1	0.3
Figure 4.7	Unrelated 2-D geometry problems	0.1	0.1	0.0	0.0	0.0	0.0	0.3
	Repetition 2-D geometry problems	0.1	0.1	0.1	0.0	0.1	0.1	0.2
	Thematically related 2-D geometry problems	0.1	0.1	0.0	0.1	0.1	0.0	0.2
	Mathematically related 2-D geometry problems	0.0	0.0	0.1	0.1	0.0	0.0	0.0
Figure 4.8	Single topic	0.1	0.0	0.0	0.0	0.1	0.1	0.0
Figure 5.1	Set-up used mathematical language or symbols only	0.0	0.0	0.0	0.0	0.0	0.0	0.0
	Set-up contained a real-life connection	0.0	0.0	0.0	0.0	0.0	0.0	0.0
Figure 5.2	Graph	0.0	0.0	0.0	0.0	0.0	0.0	0.0
	Table	0.1	0.0	0.0	0.0	0.0	0.0	0.0
	Drawing/diagram	0.1	0.0	0.0	0.1	0.0	0.0	0.1
Figure 5.3	Physical materials – problems	0.0	0.0	0.0	0.1	0.0	0.0	0.0
Figure 5.4	Physical materials – 2-D geometry problems	0.0	0.1	0.0	0.1	0.0	0.1	0.1
Figure 5.6	Applications	0.0	0.0	0.0	0.0	0.0	0.0	0.0
Figure 5.7	Target result presented publicly – concurrent problems	0.1	0.0	0.0	0.1	0.0	0.0	0.1
	Target result presented publicly – independent problems	0.0	0.0	0.0	0.0	0.0	0.0	0.0

Standard errors for estimates shown in figures and tables, by country[1]—Continued

Table/Figure	Category	AU	CZ	HK	JP[2]	NL	SW	US
Table 5.1	More than one solution method presented – problems	0.0	0.0	0.0	0.0	0.0	0.0	0.0
	More than one solution method presented – at least one	0.1	0.0	0.0	0.1	0.1	0.0	0.1
Table 5.2	Students had a choice of solution methods – problems	0.0	0.0	0.0	0.0	0.0	0.0	0.0
	Students had a choice of solution methods – at least one	0.0	0.0	0.0	0.1	0.0	0.0	0.1
Table 5.3	Examining methods – problems	0.0	0.0	0.0	0.0	0.0	0.0	0.0
	Examining methods – at least one	0.0	0.0	0.0	0.1	0.0	0.0	0.0
Table 5.4	Problem summary	0.0	0.0	0.0	0.0	0.0	0.0	0.0
Figure 5.8	Using procedures problem statement	0.1	0.0	0.0	0.1	0.1	—	0.0
	Stating concepts problem statement	0.0	0.0	0.0	0.0	0.0	—	0.0
	Making connections problem statement	0.0	0.0	0.0	0.1	0.1	—	0.0
Figure 5.9	Giving results only implementation	0.1	0.0	0.0	0.0	0.0	—	0.0
	Using procedures implementation	0.1	0.0	0.0	0.0	0.0	—	0.0
	Stating concepts implementation	0.0	0.0	0.0	0.0	0.0	—	0.0
	Making connections implementation	0.0	0.0	0.0	0.0	0.0	—	0.0
Figure 5.10	Using procedures problem statement and giving results only implementation	0.1	0.0	0.0	0.0	0.0	—	0.0
	Using procedures problem statement and using procedures implementation	0.1	0.0	0.0	0.1	0.0	—	0.0
	Using procedures problem statement and stating concepts implementation	0.0	0.0	0.0	0.1	0.0	—	0.0
	Using procedures problem statement and making connections implementation	0.0	0.0	0.0	0.0	0.0	—	0.0
Figure 5.11	Stating concepts problem statement and giving results only implementation	0.1	0.1	0.1	0.0	0.1	—	0.1
	Stating concepts problem statement and stating concepts implementation	0.1	0.1	0.1	0.0	0.1	—	0.1
	Stating concepts problem statement and making connections implementation	0.0	0.0	0.0	0.0	0.0	—	0.1

Standard errors for estimates shown in figures and tables, by country¹—Continued

Table/Figure	Category	AU	CZ	HK	JP²	NL	SW	US
Figure 5.12	Making connections problem statement and giving results only implementation	0.1	0.0	0.0	0.0	0.0	—	0.1
	Making connections problem statement and using procedures implementation	0.1	0.1	0.1	0.1	0.1	—	0.1
	Making connections problem statement and stating concepts implementation	0.1	0.0	0.1	0.0	0.1	—	0.0
	Making connections problem statement and making connections implementation	0.0	0.1	0.1	0.0	0.1	—	0.0
Figure 5.13	Repeating procedures assignment	0.1	0.0	0.0	0.0	0.1	0.0	0.0
	Other than repeating procedures assignment or mix	0.0	0.0	0.0	0.0	0.0	0.0	0.0
Table 5.5	Mathematical information	0.0	0.0	0.0	0.0	0.1	0.0	0.0
	Contextual information	0.1	0.0	0.0	0.0	0.1	0.0	0.0
	Mathematical activity	0.0	0.0	0.0	0.0	0.0	0.0	0.0
	Announcements	0.0	0.0	0.0	0.0	0.1	0.0	0.0
Figure 5.14	Student words per 50 minutes of public interaction	55.4	35.4	59.1	89.9	85.9	—	71.7
	Teacher words per 50 minutes of public interaction	143.2	98.9	143.2	99.4	165.0	—	166.4
Figure 5.15	Teacher words to every one student word	0.8	0.6	1.4	2.1	1.7	—	0.6
Figure 5.16	Teacher utterances that were 25+ words	0.0	0.0	0.0	0.0	0.0	0.0	0.0
	Teacher utterances that were 5+ words	0.0	0.0	0.0	0.0	0.0	0.0	0.0
	Teacher utterances that were 1–4 words	0.0	0.0	0.0	0.0	0.0	0.0	0.0
Figure 5.17	Student utterances that were 10+ words	0.0	0.0	0.0	0.0	0.0	—	0.0
	Student utterances that were 5+ words	0.0	0.0	0.0	0.0	0.0	—	0.0
	Student utterances that were 1–4 words	0.0	0.0	0.0	0.0	0.0	—	0.0
Table 5.6	Chalkboard	0.0	0.0	0.0	0.0	0.0	0.0	0.1
	Projector	0.1	0.0	0.0	0.0	0.0	0.0	0.1
	Textbook or worksheet	0.0	0.0	0.0	0.0	0.0	0.0	0.0
	Special mathematics materials	0.1	0.1	0.0	0.0	0.0	0.0	0.1
	Real-world objects	0.0	0.0	0.0	0.1	0.0	0.0	0.0
Figure 5.18	Computational calculators	0.1	0.0	0.1	0.0	0.0	0.0	0.1

Appendix C
Standard Errors for Estimates Shown in Figures and Tables

Standard errors for estimates shown in figures and tables, by country[1]—Continued

Table/Figure	Category	AU	CZ	HK	JP2	NL	SW	US
Figure D.1	Elementary	0.1	0.0	0.0	—	0.1	0.1	0.1
	Elementary/moderate	0.1	0.1	0.1	—	0.1	0.1	0.1
	Moderate	0.1	0.1	0.1	—	0.1	0.1	0.1
	Moderate/advanced	0.1	0.1	0.1	—	0.1	0.1	0.1
	Advanced	0.0	0.1	0.1	—	0.0	0.1	0.0
Table D.1	Conceptual	0.1	0.1	0.1	—	0.1	0.1	0.1
	Procedural	0.0	0.0	0.0	—	0.0	0.0	0.1
	Notational	0.1	0.1	0.1	—	0.1	0.1	0.1
Figure D.2	Deductive reasoning	0.0	0.1	0.1	—	0.1	0.1	0.1
Figure D.3	Development of a rationale	0.1	0.1	0.1	—	0.1	0.1	0.0
Figure D.4	Generalizations	0.1	0.1	0.1	—	0.1	0.1	0.0
Figure D.5	Fragmented	0.0	0.1	0.0	—	0.1	0.0	0.0
	Moderately fragmented	0.1	0.1	0.0	—	0.1	0.1	0.1
	Mixed	0.1	0.1	0.0	—	0.1	0.1	0.1
	Moderately thematic	0.1	0.1	0.1	—	0.1	0.1	0.1
	Thematic	0.1	0.1	0.1	—	0.1	0.1	0.1
Figure D.6	Undeveloped	0.1	0.0	0.0	—	0.1	0.1	0.1
	Partially developed	0.1	0.1	0.1	—	0.1	0.1	0.1
	Moderately developed	0.1	0.1	0.1	—	0.1	0.1	0.1
	Substantially developed	0.1	0.1	0.1	—	0.1	0.1	0.1
	Fully developed	0.0	0.1	0.1	—	0.1	0.1	0.1
Figure D.7	Very unlikely	0.0	0.1	0.0	—	0.1	0.1	0.1
	Doubtful	0.1	0.1	0.1	—	0.1	0.1	0.1
	Possible	0.1	0.1	0.1	—	0.1	0.1	0.1
	Probable	0.1	0.1	0.1	—	0.1	0.1	0.1
	Very likely	0.1	0.1	0.1	—	0.1	0.1	0.0
Figure D.8	Low	0.1	0.1	0.0	—	0.1	0.1	0.1
	Moderately low	0.1	0.1	0.1	—	0.1	0.1	0.1
	Moderate	0.1	0.1	0.1	—	0.1	0.1	0.1
	Moderately high	0.1	0.1	0.1	—	0.1	0.1	0.1
	High	0.1	0.1	0.1	—	0.1	0.1	0.0

Standard errors for estimates shown in figures and tables, by country[1]—Continued

Table/Figure	Category	AU	CZ	HK	JP[2]	NL	SW	US
Figure D.9	Coherence	0.3	0.3	0.1	—	0.3	0.3	0.3
	Presentation	0.2	0.2	0.2	—	0.2	0.3	0.3
	Student engagement	0.2	0.2	0.2	—	0.3	0.3	0.2
	Overall quality	0.3	0.3	0.2	—	0.3	0.3	0.3

—Not available. Country was not included in analysis.
[1] AU=Australia; CZ=Czech Republic; HK=Hong Kong SAR; NL=Netherlands; SW=Switzerland; and US=United States.
[2] Japanese mathematics data were collected in 1995.
SOURCE: U.S. Department of Education, National Center for Education Statistics, Third International Mathematics and Science Study (TIMSS), Video Study, 1999.

APPENDIX D
Results From the Mathematics Quality Analysis Group

Judgments about the mathematics content of the lessons were made by the mathematics quality analysis group, a team of mathematicians and teachers of post-secondary mathematics (see appendix A for a description of this group). The mathematics quality analysis group developed a coding scheme that focused on the content of the lessons and then applied the scheme by reaching consensus on each judgment. The members of this group examined country-blind written records of 20 lessons selected randomly from each country's sample (see appendix A). Japan was not included because the same group already had analyzed Japanese lessons as part of the TIMSS 1995 Video Study which meant, among other things, that potential country bias could not be adequately reduced (see Stigler et al. [1999] and Manaster [1998] for a report of the group's findings in the 1995 study).

The findings of the mathematics quality analysis group are based on a relatively small, randomly selected sub-sample of lessons and, consequently, are considered preliminary. Because the results are based on a sub-sample, the results are raw percentages rather than weighted percentages. The percentages and ratings shown in the figures are descriptive only; no statistical comparisons were made. Readers are urged to be cautious in their interpretations of these results because the sub-sample, due to its relatively small size, might not be representative of the entire sample or of eighth-grade mathematics lessons in each country.

Curricular Level of the Content

The data shown in chapter 4, table 4.1, display the relative emphasis given to different topics, on average, across the full sample in each country. These percentages provide one estimate of the level of content. Another estimate can be obtained by asking experts in the field to review the lessons for the curricular level of the content.

One of the codes developed by the mathematics quality analysis group placed each lesson in the sub-sample into one of five curricular levels, from elementary (1) to advanced (5). The moderate or mid level (3) was defined to include content that usually is encountered by students just prior to the standard topics of a beginning algebra course that often is taught in the eighth grade. One rating was assigned to each lesson based on the rating that best described the content of the lesson, taken as a whole.

Figure D.1 shows the percentage of eighth-grade mathematics lessons assigned to each rating. Because these analyses were limited to a subset of the total sample of lessons, the percentages were not compared statistically and the results should be interpreted with caution. This same cautionary note applies to all of the findings of the mathematics quality analysis group and is noted on each of the figures that present the findings of this group. In figure D.1, as in all figures in this special section, percentages indicate the number of lessons in the sub-sample that contain a particular feature. In other words, 100 percent and 0 percent are indications that all or none of the sub-sample lessons contained a feature, and are not meant to imply that all or no lessons in the country contain the feature.

| FIGURE D.1. | Percentage of eighth-grade mathematics lessons in sub-sample at each content level, by country: 1999 |

[Stacked bar chart showing percentage of sub-sampled lessons by content level for each country]

Legend:
- Advanced (5)
- Moderate/advanced (4)
- Moderate (3)
- Elementary/moderate (2)
- Elementary (1)

Values by country:
- AU: Elementary 10, Elementary/moderate 45, Moderate 30, Moderate/advanced 15, Advanced 0
- CZ: Elementary 15, Elementary/moderate 20, Moderate 45, Moderate/advanced 20, Advanced 0
- HK: Elementary 5, Elementary/moderate 40, Moderate 35, Moderate/advanced 20, Advanced 0
- NL: Elementary 10, Elementary/moderate 20, Moderate 40, Moderate/advanced 30, Advanced 0
- SW: Elementary 15, Elementary/moderate 15, Moderate 30, Moderate/advanced 35, Advanced 5
- US: Elementary 15, Elementary/moderate 25, Moderate 40, Moderate/advanced 20, Advanced 0

Country[1]

[1]AU=Australia; CZ=Czech Republic; HK=Hong Kong SAR; NL=Netherlands; SW=Switzerland; and US=United States.
NOTE: Lessons included here are a random sub-sample of lessons in each country. Results should be interpreted with caution because they might not be representative of the entire sample. The number in the parentheses is the ranking number for that category. A moderate ranking was defined to include content that usually is encountered by students just prior to the standard topics of a beginning algebra course that is often taught in the eighth grade.
SOURCE: U.S. Department of Education, National Center for Education Statistics, Third International Mathematics and Science Study (TIMSS), Video Study, 1999.

Averaging the content level ratings of each country's sub-sample of lessons gives a summary rating for each country. Additional caution is needed when interpreting the summary ratings because, for example, an elementary lesson and an advanced lesson are unlikely to average to the same experience for students as two moderate lessons. With this caveat in mind, the ratings for countries with the most advanced (5) to the most elementary (1) content in the sub-sample of lessons, were the Czech Republic and Hong Kong SAR (3.7), Switzerland (3.0), the Netherlands (2.9), the United States (2.7), and Australia (2.5) (see also figure D.9).

Nature of the Content

The distinction between different kinds of mathematical knowledge has been used frequently by researchers to describe different kinds of mathematics learning and to describe the outcomes of different kinds of learning environments (Hiebert 1986). Common distinctions separate knowledge of concepts, procedures, and written notation or definitions.

The mathematics quality analysis group characterized the mathematics presented in the sub-sample of lessons as conceptual, procedural, or notational. Conceptual mathematics was defined as the development of mathematical ideas or procedures. Segments of conceptual mathematics might include examples and explanations for why things work like they do. Procedural mathematics was defined as the presentation of mathematical procedures without much explanation, or the practice of procedures that appeared to be known already by the students.

Often, the development and first application of a solution procedure was coded as conceptual whereas subsequent applications of the method were coded as procedural. The notational code was used when the presentation or discussion centered on mathematical definitions or notational conventions.

Table D.1 shows the percentage of eighth-grade mathematics lessons that contained segments of conceptual, procedural, and notational mathematics. Because these analyses were limited to a subset of the total sample of lessons, the percentages were not compared statistically and the results should be interpreted with caution.

TABLE D.1. Percentage of eighth-grade mathematics lessons in sub-sample that contained segments of conceptual, procedural, and notational mathematics, by country: 1999

Country	Conceptual	Procedural	Notational
Australia	55	100	35
Czech Republic	60	100	40
Hong Kong SAR	50	100	55
Netherlands	40	100	45
Switzerland	75	100	45
United States	45	95	45

Percentage of sub-sampled lessons that contained segments of the following types of mathematics:

NOTE: Lessons included here are a random sub-sample of lessons in each country. Results should be interpreted with caution because they might not be representative of the entire sample.
SOURCE: U.S. Department of Education, National Center for Education Statistics, Third International Mathematics and Science Study (TIMSS), Video Study, 1999.

Almost all lessons in the sub-samples for every country contained segments that were coded procedural. Between 35 percent and 55 percent of the lessons in the sub-samples contained segments of notational mathematics and between 40 percent and 75 percent of the lessons in the sub-samples contained segments of conceptual mathematics.

Mathematical Reasoning

As noted in chapter 4, one hallmark of doing mathematics is engaging in special forms of reasoning, such as deduction (deriving conclusions from stated assumptions using a logical chain of inferences). Other forms of mathematical reasoning include generalization (recognizing that several examples share more general properties) and using counter-examples (finding one example that does not work to prove that a mathematical conjecture cannot be true). These special reasoning processes provide one way to distinguish mathematics from other disciplines (National Research Council 2001a; Whitehead 1948).

Because the findings from the TIMSS 1995 Video Study indicated that not all countries provide eighth-graders opportunities to engage in deductive reasoning (Stigler et al. 1999; Manaster 1998), the mathematics quality analysis group expanded the mathematical reasoning coding scheme it had used for the 1995 Video Study in an attempt to identify special reasoning forms that might be present in eighth-grade mathematics lessons.

Figure D.2 shows the results of applying the group's definition of deductive reasoning to the sub-sample of eighth-grade mathematics lessons. Such reasoning could occur as part of problem or non-problem segments. The reasoning did not need to include a formal proof, only a logical chain of inferences with some explanation.

The percentage of eighth-grade mathematics lessons in the sub-sample that contained deductive reasoning by the teacher or students is shown in figure D.2. Because these analyses were limited to a subset of the total sample of lessons, the percentages were not compared statistically and the results should be interpreted with caution.

FIGURE D.2. Percentage of eighth-grade mathematics lessons in sub-sample that contained deductive reasoning, by country: 1999

Country	Percentage
AU	0
CZ	5
HK	15
NL	5
SW	10
US	10

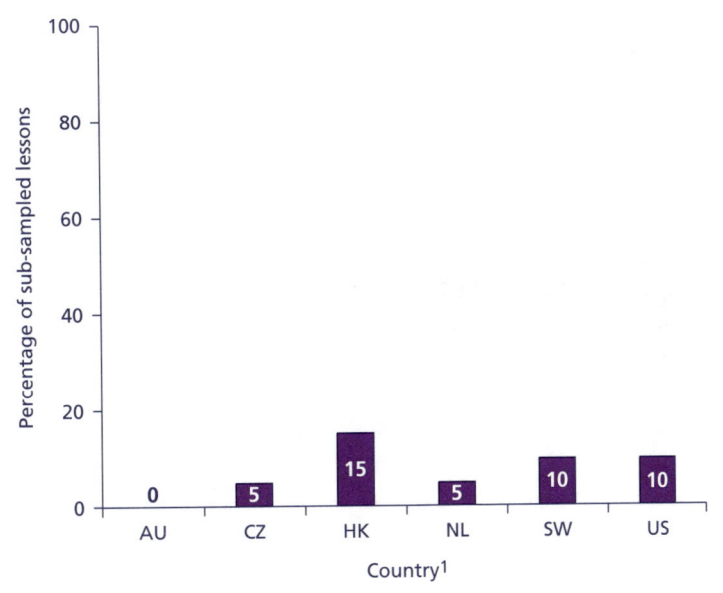

[1] AU=Australia; CZ=Czech Republic; HK=Hong Kong SAR; NL=Netherlands; SW=Switzerland; and US=United States.
NOTE: Lessons included here are a random sub-sample of lessons in each country. Results should be interpreted with caution because they might not be representative of the entire sample.
SOURCE: U.S. Department of Education, National Center for Education Statistics, Third International Mathematics and Science Study (TIMSS), Video Study, 1999.

A maximum of 15 percent of the sub-samples of lessons in any country contained instances of deductive reasoning. As noted earlier, Japanese lessons were not included in these analyses because the mathematics quality analysis group had examined a sub-sample of Japanese lessons for the TIMSS 1995 Video Study.

Deductive reasoning is not the only special form of mathematical reasoning. The mathematics quality analysis group coded the previously described sub-sample of 20 lessons in each country (except Japan) for other special kinds of mathematical reasoning in which eighth-graders seem to be capable of engaging (National Research Council 2001a). "Developing a rationale" was defined by the mathematics quality analysis group as explaining or motivating, in broad mathematical terms, a mathematical assertion or procedure. For example, teachers might show that the rules for adding and subtracting integers are logical extensions of those for adding and subtracting whole numbers, and that these more general rules work for all numbers. When such explanations took a systematic logical form, they were coded as deductive reasoning (see figure D.2); when they took a less systematic or precise form, they were coded as developing a rationale.

Figure D.3 shows that a maximum of 25 percent of the eighth-grade mathematics lessons in any country's sub-sample included instances of developing a rationale. As before, these analyses were limited to a subset of the total sample of lessons so the percentages were not compared statistically and the results should be interpreted with caution.

FIGURE D.3. Percentage of eighth-grade mathematics lessons in sub-sample that contained the development of a rationale, by country: 1999

Country	Percentage
AU	25
CZ	10
HK	20
NL	10
SW	25
US	0

[1]AU=Australia; CZ=Czech Republic; HK=Hong Kong SAR; NL= Netherlands; SW=Switzerland; and US=United States.
NOTE: Lessons included here are a random sub-sample of lessons in each country. Results should be interpreted with caution because they might not be representative of the entire sample.
SOURCE: U.S. Department of Education, National Center for Education Statistics, Third International Mathematics and Science Study (TIMSS), Video Study, 1999.

The mathematics quality analysis group also examined the sub-sample of lessons to determine the occurrence of two other forms of reasoning: generalization and counter-example. Generalization might involve, for example, graphing several linear equations such as $y = 2x + 3$,

2y = x - 2, and y = -4x, and making an assertion about the role played by the numbers in these equations in determining the position and slope of the associated lines. Generalization, then, involves inducing general properties or principles from several examples.

As shown in figure D.4, generalizations occurred in a maximum of 20 percent of the eighth-grade mathematics lessons in any country. Because these analyses were limited to a subset of the total sample of lessons, the percentages were not compared statistically and the results should be interpreted with caution.

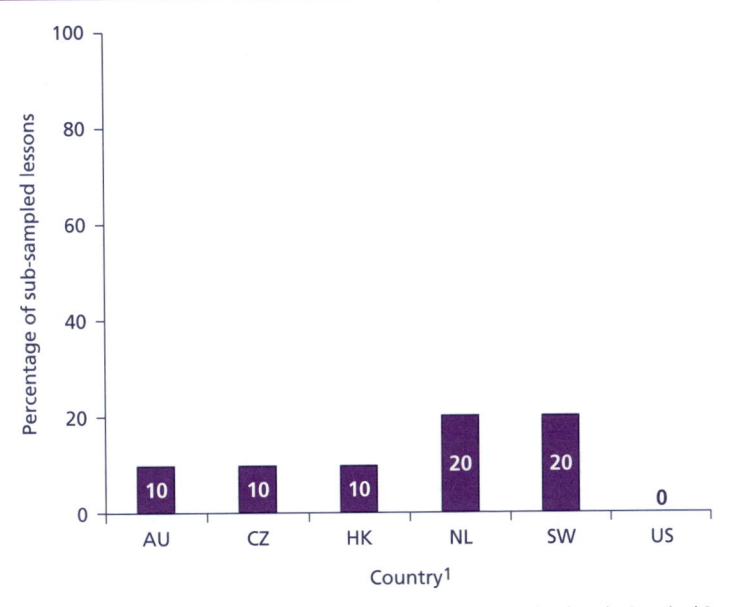

FIGURE D.4. Percentage of eighth-grade mathematics lessons in sub-sample that contained generalizations, by country: 1999

Country[1]	Percentage
AU	10
CZ	10
HK	10
NL	20
SW	20
US	0

[1] AU=Australia; CZ=Czech Republic; HK=Hong Kong SAR; NL= Netherlands; SW=Switzerland; and US=United States.
NOTE: Lessons included here are a random sub-sample of lessons in each country. Results should be interpreted with caution because they might not be representative of the entire sample.
SOURCE: U.S. Department of Education, National Center for Education Statistics, Third International Mathematics and Science Study (TIMSS), Video Study, 1999.

A final kind of special mathematical reasoning—using a counter-example—involves finding an example to show that an assertion cannot be true. For instance, suppose someone claims that the area of a rectangle gets larger whenever the perimeter gets larger. A counter-example would be a rectangle whose perimeter becomes larger but the area does not become larger.

The mathematics quality analysis group found that, in the sub-sample of eighth-grade mathematics lessons, demonstrating that a conjecture cannot be true by showing a counter-example occurred in 10 percent of the lessons in Australia and 5 percent of the lessons in Hong Kong SAR. The other countries showed no evidence of counter-example use.

Overall Judgments of Mathematical Quality

The mathematics quality analysis group judged the overall quality of the mathematics in the sub-sample of lessons along several dimensions: coherence, presentation, student engagement, and overall quality. Each lesson was rated from 1 (low) to 5 (high) on each dimension. Whereas most of the group's codes reported to this point marked the occurrence of particular features of a lesson, the group's overall judgments of quality considered each lesson as a whole. As stated earlier, country-identifying marks had been removed from the written records to mask the country from which the lessons came. Recall also that Japanese lessons were not included in the group's sub-sample.

Coherence was defined by the group as the (implicit and explicit) interrelation of all mathematical components of the lesson. A rating of 1 indicated a lesson with multiple unrelated themes or topics and a rating of 5 indicated a lesson with a central theme that progressed saliently through the whole lesson.

Figure D.5 shows the percentage of eighth-grade mathematics lessons in the sub-sample assigned to each level of coherence. Averaging across all the lessons within each country's sub-sample yields the following general ratings of countries based on lesson coherence: Hong Kong SAR (4.9), Switzerland (4.3), Australia (4.2), the Netherlands (4.0), the Czech Republic (3.6), and the United States (3.5). Because these analyses were limited to a subset of the total sample of lessons, the ratings were not compared statistically and the results should be interpreted with caution.

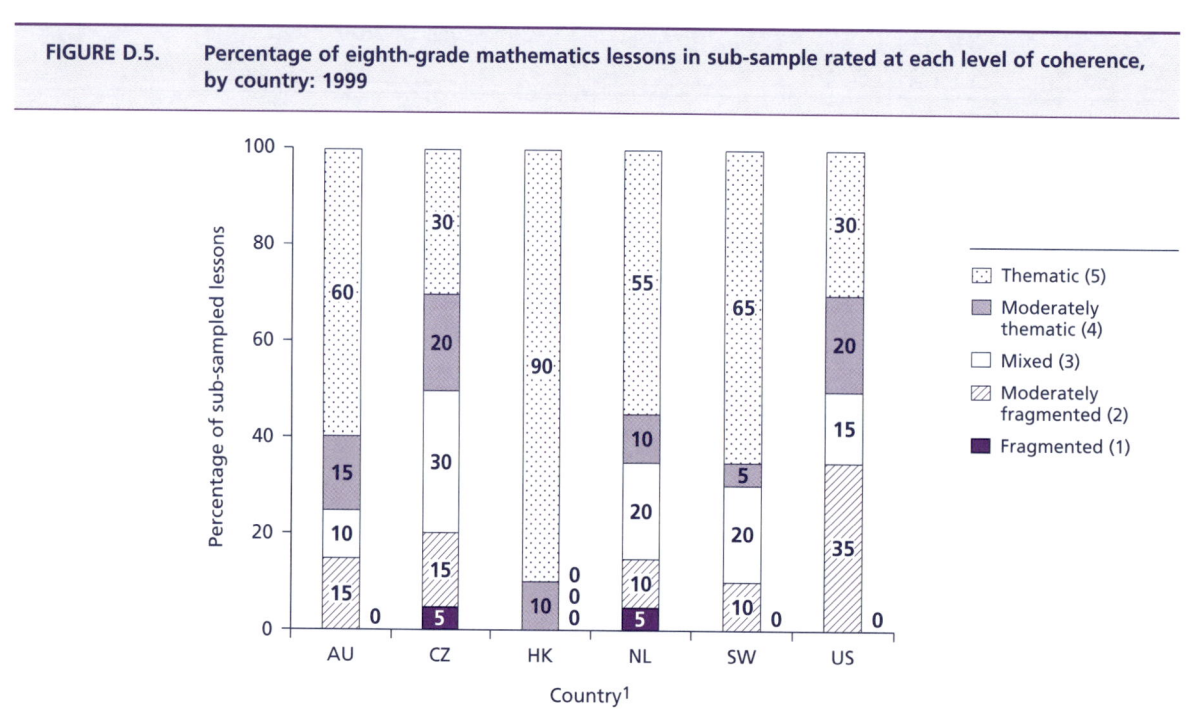

FIGURE D.5. Percentage of eighth-grade mathematics lessons in sub-sample rated at each level of coherence, by country: 1999

[1]AU=Australia; CZ=Czech Republic; HK=Hong Kong SAR; NL=Netherlands; SW=Switzerland; and US=United States.
NOTE: Lessons included here are a random sub-sample of lessons in each country. Results should be interpreted with caution because they might not be representative of the entire sample. The number in the parentheses is the ranking number for that category. For Hong Kong SAR, no lessons in the sub-sample were found to be mixed, moderately fragmented, or fragmented.
SOURCE: U.S. Department of Education, National Center for Education Statistics, Third International Mathematics and Science Study (TIMSS), Video Study, 1999.

It is worth pausing here to make an observation about the results related to lesson coherence. The Czech Republic makes an interesting case. The full sample of eighth-grade mathematics lessons from the Czech Republic contained, on average, a higher percentage of unrelated problems per lesson than Hong Kong SAR, the Netherlands, and Switzerland (see figure 4.6 in chapter 4), and the mathematics quality analysis group judged that 50 percent of their sub-sample of Czech lessons contained at least moderately fragmented portions (figure D.5). But recall the findings in chapter 3, which showed that the lessons in the Czech Republic displayed relatively high profiles of pedagogical coherence (e.g., lesson goal and summary statements, and interruptions to lessons), compared with some of the other countries (figures 3.12, 3.13, 3.15, and 3.16). This suggests that there are several dimensions of lesson coherence and that they are not necessarily interdependent.

Another characteristic of overall quality defined by the mathematics quality analysis group was presentation—the extent to which the lesson included some development of the mathematical concepts or procedures. Development required that mathematical reasons or justifications were given for the mathematical results presented or used. This might be done, for example, by the teacher drawing clear connections between what was known and familiar to the students and what was unknown. Presentation ratings took into account the quality of mathematical arguments. Higher ratings meant that sound mathematical reasons were provided by the teacher (or students) for concepts and procedures. Mathematical errors made by the teacher reduced the ratings. A rating of 1 indicated a lesson that was descriptive or routinely algorithmic with little mathematical justification provided by the teacher or students for why things work like they do. A rating of 5 indicated a lesson in which the concepts and procedures were mathematically motivated, supported, and justified by the teacher or students.

Figure D.6 shows the percentage of eighth-grade mathematics lessons in the sub-sample assigned to each level of presentation. Averaging the ratings of all the lessons within each country's sub-sample yields the following general ratings of countries based on presentation: Hong Kong SAR (3.9), Switzerland (3.4), the Czech Republic (3.3), Australia (3.0), the Netherlands (2.8), and the United States (2.4). Because these analyses were limited to a subset of the total sample of lessons, the ratings were not compared statistically and the results should be interpreted with caution.

FIGURE D.6. Percentage of eighth-grade mathematics lessons in sub-sample rated at each level of presentation, by country: 1999

[Stacked bar chart showing percentages for AU, CZ, HK, NL, SW, US across five categories: Fully developed (5), Substantially developed (4), Moderately developed (3), Partially developed (2), Undeveloped (1).]

Category	AU	CZ	HK	NL	SW	US
Fully developed (5)	0	10	20	5	15	5
Substantially developed (4)	40	30	55	20	25	10
Moderately developed (3)	30	40	—	30	45	20
Partially developed (2)	20	20	15	35	5	40
Undeveloped (1)	10	0	10	10	15	—

[1]AU=Australia; CZ=Czech Republic; HK=Hong Kong SAR; NL=Netherlands; SW=Switzerland; and US=United States.
NOTE: Lessons included here are a random sub-sample of lessons in each country. Results should be interpreted with caution because they might not be representative of the entire sample. The number in the parentheses is the ranking number for that category.
SOURCE: U.S. Department of Education, National Center for Education Statistics, Third International Mathematics and Science Study (TIMSS), Video Study, 1999.

Student engagement was defined by the mathematics quality analysis group as the likelihood that students would be actively engaged in meaningful mathematics during the lesson. A rating of very unlikely (1) indicated a lesson in which students were asked to work on few of the problems in the lesson and those problems did not appear to stimulate reflection on mathematical concepts or procedures. In contrast, a rating of very likely (5) indicated a lesson in which students were expected to work actively on, and make progress solving, problems that appeared to raise interesting mathematical questions for them and then to discuss their solutions with the class.

Figure D.7 shows the percentage of eighth-grade mathematics lessons in the sub-sample assigned to each level of student engagement. Averaging across all the lessons within each country's sub-sample yields the following general ratings of countries based on student engagement: Hong Kong SAR (4.0), the Czech Republic (3.6), Switzerland (3.3), Australia (3.2), the Netherlands (2.9), and the United States (2.4). Because these analyses were limited to a subset of the total sample of lessons, the ratings were not compared statistically and the results should be interpreted with caution.

Appendix D
Results From the Mathematics Quality Analysis Group

FIGURE D.7. Percentage of eighth-grade mathematics lessons in sub-sample rated at each level of student engagement, by country: 1999

Rating	AU	CZ	HK	NL	SW	US
Very likely (5)	10	10	35	10	10	0
Probable (4)	30	55	30	20	40	15
Possible (3)	30	20	30	30	30	25
Doubtful (2)	30	10	5	30	10	45
Very unlikely (1)	0	5	0	10	10	15

[1]AU=Australia; CZ=Czech Republic; HK=Hong Kong SAR; NL=Netherlands; SW=Switzerland; and US=United States.
NOTE: Lessons included here are a random sub-sample of lessons in each country. Results should be interpreted with caution because they might not be representative of the entire sample. The number in the parentheses is the ranking number for that category.
SOURCE: U.S. Department of Education, National Center for Education Statistics, Third International Mathematics and Science Study (TIMSS), Video Study, 1999.

The mathematics quality analysis group made a final overall judgment about the quality of mathematics in each lesson in their sub-sample. This overall quality judgment took into account all previous codes and was defined as the opportunities that the lesson provided for students to construct important mathematical understandings. Ratings ranged from 1 for low to 5 for high.

Figure D.8 shows the percentage of eighth-grade mathematics lessons in the sub-sample assigned to each level of overall quality. Averaging across all the lessons within each country's sub-sample yields the following general rating of countries based on overall quality of the mathematics presented: Hong Kong SAR (4.0), the Czech Republic (3.4), Switzerland (3.3), Australia (2.9), the Netherlands (2.7), and the United States (2.3). Because these analyses were limited to a subset of the total sample of lessons, the ratings were not compared statistically and the results should be interpreted with caution.

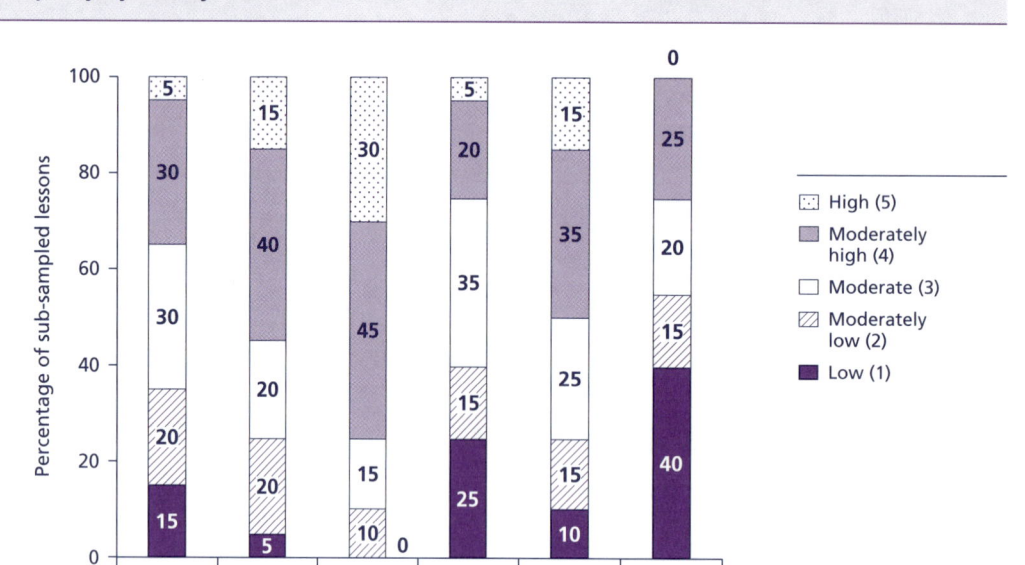

FIGURE D.8. Percentage of eighth-grade mathematics lessons in sub-sample rated at each level of overall quality, by country: 1999

[Stacked bar chart showing percentages by country (AU, CZ, HK, NL, SW, US) across five quality levels: High (5), Moderately high (4), Moderate (3), Moderately low (2), Low (1).

AU: Low 15, Mod. low 20, Moderate 30, Mod. high 30, High 5
CZ: Low 5, Mod. low 20, Moderate 20, Mod. high 40, High 15
HK: Low 0, Mod. low 10, Moderate 15, Mod. high 45, High 30
NL: Low 25, Mod. low 15, Moderate 35, Mod. high 20, High 5
SW: Low 10, Mod. low 15, Moderate 25, Mod. high 35, High 15
US: Low 40, Mod. low 15, Moderate 20, Mod. high 25, High 0]

[1]AU=Australia; CZ=Czech Republic; HK=Hong Kong SAR; NL=Netherlands; SW=Switzerland; and US=United States.
NOTE: Lessons included here are a random sub-sample of lessons in each country. Results should be interpreted with caution because they might not be representative of the entire sample. The number in the parentheses is the ranking number for that category.
SOURCE: U.S. Department of Education, National Center for Education Statistics, Third International Mathematics and Science Study (TIMSS), Video Study, 1999.

A summary display of the overall judgments of the mathematics quality analysis group is found in figure D.9. The general ratings reported above for the sub-sample of lessons of each country for coherence, presentation, student engagement, and overall mathematics quality are plotted on the same figure. As the figure shows, the relative standing of Hong Kong SAR was consistently high and the relative standing of the United States was consistently low. The other four countries received general ratings that fell in between and that varied depending on the dimension examined. Again, these ratings were based on a sub-sample of lessons and, therefore, might not be representative of the entire sample and of eighth-grade mathematics lessons in each country.

FIGURE D.9. General ratings of the sub-sample of eighth-grade mathematics lessons for each overall dimension of content quality, by country: 1999

Dimension	Australia ◆	Czech Republic ■	Hong Kong SAR ▲	Netherlands ×	Switzerland ○	United States ●
Coherence	4.2	3.6	4.9	4.0	4.3	3.5
Presentation	3.0	3.3	3.9	2.8	3.4	2.4
Student engagement	3.2	3.6	4.0	2.9	3.3	2.4
Overall quality	2.9	3.4	4.0	2.7	3.3	2.3

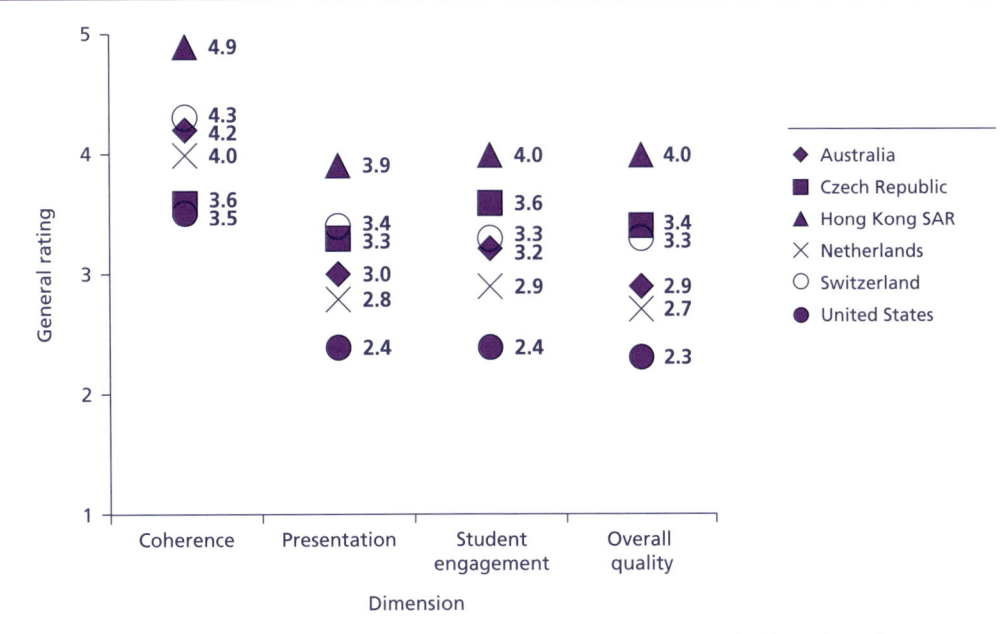

NOTE: Lessons included here are a random sub-sample of lessons in each country. Results should be interpreted with caution as they may not be representative of the entire sample. Rating along each dimension based on scale of 1 to 5, with 1 being the lowest possible rating and 5 being the highest.
SOURCE: U.S. Department of Education, National Center for Education Statistics, Third International Mathematics and Science Study (TIMSS), Video Study, 1999.

Summary

It is difficult to draw conclusions about similarities and differences among countries from the findings just presented because of their descriptive and exploratory character. Two points are worth noting, however.

A first point is that where there is overlap between the variables defined by the mathematics quality analysis group and those described in the chapters of this report, the findings are not inconsistent with each other. For example, in chapter 4 it was reported that a relatively small percentage of mathematics problems (and lessons) in countries other than Japan involved proofs. The mathematics quality analysis group found similarly infrequent instances of special mathematical reasoning, including deductive reasoning.

A second point is that the findings reported in this appendix could be considered hypotheses worthy of further examination. Because the quality of mathematical content is theoretically an important contributor to the learning opportunities for students (National Research Council 2001a), and because the mathematics quality analysis group developed a series of high-inference codes for evaluating the quality of content, it is likely that an application of the codes to the full sample of TIMSS 1999 Video Study lessons would add to the findings presented in chapter 4. In addition, the results presented in this appendix identify constructs of mathematical content that would benefit from development and further application in other studies that aim to describe the quality of content in mathematics lessons.

APPENDIX E
Hypothesized Country Models

Hypotheses were developed about specific instructional patterns that might be found in eighth-grade mathematics classrooms in each country.[1] The process began by considering the four or five field test videos collected in each country—eighth-grade mathematics lessons that provided an initial opportunity to observe teaching in the different countries in the sample.[2] An international group of representatives (i.e., the field test team) met together for an entire summer, viewing and reflecting on these tapes. They followed a structured protocol to generate hypotheses that could later be tested by quantitative analyses of the full data set. First, the country representatives closely examined field test lessons from their own country, and nominated the one that was "most typical." Then, the entire group viewed and discussed each typical lesson at length, noting in particular the similarities and differences among countries. These discussions provided consensus that six dimensions framed mathematics classroom practice and were of interest across countries and lessons: Purpose, Classroom Routine, Actions of Participants, Content, Classroom Talk, and Climate. These dimensions were then used to create hypothesized country models—holistic representations of a "typical" mathematics lesson in each country.

The hypothesized country models were presented to National Research Coordinators, the Mathematics Steering Committee, and other colleagues in each country including eighth-grade mathematics teachers and educators, and refined over a period of several months. The goal was to retain an "insider perspective," and faithfully represent in the coding system the critical features of eighth-grade mathematics teaching in each country. The hypothesized country models served two purposes toward this end. First, the models provided a basis on which to identify key, universal variables for quantitative coding. Second, they described a larger context that might be useful in interpreting the coding results. The hypothesized country models are presented in this appendix.

TABLE E.1. Key to symbols and acronyms used in hypothesized models

Symbol/acronym	Meaning
T	Teacher
S	Student
Ss	Students
HW	Homework
BB	Blackboard
‖: :‖	Segment may repeat

---

[1] The process of creating a hypothesized country model was not completed for Japan.
[2] Field test lessons were not collected in Hong Kong SAR because a final decision about participation in the study had not yet been made.

Appendix E | 205
Hypothesized Country Models

FIGURE E.1. Hypothesized country model for Australia

Purpose	Review	Introduction of new material	Assignment of task	Practice/application and re-instruction			Conclusion
				Practice/application	Reassignment of task	Practice/application	
	Reinforce knowledge; check/correct/review homework; re-instruct	Acquisition of knowledge	Assignment of task	Application of knowledge	Assignment of task	Application of knowledge	Reinforce knowledge
Classroom routine	review of relevant material previously worked on	presentation of new material	assignment of task	completion of task	assignment of task	completion of task	summary of new material; assignment of homework
Actions of participants	T – [at front] ask Ss questions; elicit/embellish responses; demonstrate examples on BB	T – [at front] provides information asking some Ss questions and using examples on BB	T – [at front] describes textbook/worksheet task	T – [roams room] provides assistance to Ss as needed and observes Ss progress on set task	T – [at front] re-explains textbook/worksheet task	T – [roams room] provides assistance to Ss as needed and observes Ss progress on set task	T – [at front] provides information and asks Ss questions
	Ss – [in seats] respond to and ask T questions; listen to T explanations, watch demonstrations	Ss – [in seats] listen to T explanations and respond to T questions	Ss – [in seats] listen to T descriptions	Ss – [in seats] work individually or in pairs on task	Ss – [in seats] listen to T descriptions	Ss – [in seats] work individually or in pairs on task	Ss – [in seats] listen to T descriptions; respond to and ask T questions
Content	related to previous lesson	definitions/examples building on ideas previously worked on	description of task; focus on textbook/worksheet problems	textbook/worksheet problems	description of task; focus on textbook/worksheet problems	textbook/worksheet problems	textbook/worksheet problems; homework problems
Classroom talk	T talks most; Ss one-word responses	Mix of T/S talk although discussion clearly T directed	T provides direct instructions	mix of T/S and S/S talk – including explanations and questions	T provides direct instructions	mix of T/S and S/S talk – including explanations and questions	mix of T and T/Ss talk – including explanations and questions
Climate	somewhat informal – relaxed yet focused						

SOURCE: U.S. Department of Education, National Center for Education Statistics, Third International Mathematics and Science Study (TIMSS), Video Study, 1999.

FIGURE E.2. Hypothesized country model for the Czech Republic

Purpose	Review			Constructing new knowledge				Practice	
	Evaluating	Securing old knowledge	Re-instruction	Activating old knowledge	Constructing new topics	Formulating the new information		Practice "working-through"	Using knowledge in different problems
Classroom routine	oral exam test homework	set of problems; homework	dialogue	Experiment, solving problems, demonstration, dialogue	dialogue			dialogue solving problems	solving problems
Actions of participants	T – giving grade	T – gives individual help	T – explaining procedure			T – writing notes at the board			
	Ss – solving problems at the board	Ss – at the board		Ss – answering questions, solving problems at the board				Ss – solving problem; more then one student solving one problem	
Classroom talk	answering questions; fast pace		T talk most	T-S dialogue	teacher talks most of the time; slow pace	teacher talks most of the time			more mathematically open questions
Content	content probably from unit			special problems prepared in special order, solutions are very visible, strong connection with new topics	step-by-step solving problem, solutions very visible	mathematical statements and definitions; something new that students don't know			stronger connection with real life
Climate	few mistakes allowed students very quiet serious atmosphere	mistakes are not graded but not expected, students talk loudly						more mistakes allowed	

SOURCE: U.S. Department of Education, National Center for Education Statistics, Third International Mathematics and Science Study (TIMSS), Video Study, 1999.

Appendix E | **Hypothesized Country Models**

FIGURE E.3. Hypothesized country model for Hong Kong SAR

Purpose	Review	Instruction	Consolidation	
	To review material learned in the past To prepare for the present lesson	To introduce and explain new concepts and/or skills	To practice the skills learned	
Classroom routine	T – goes over relevant material learned in the past, sometimes through asking Ss questions	T – introduces a new topic T – explains the new concepts/skills T – shows one or more worked examples	Seatwork T – assigns seatwork Ss – work on seatwork T – helps individual Ss	Evaluation T – asks some Ss to work on the board T – discusses the work on the board with Ss	Homework T – assigns homework Ss – start doing homework
Actions of participants	T – talks at the blackboard Ss – listen in their seats Ss – answer questions from their seats	T – explains at the blackboard T – works examples on the blackboard Ss – listen and/or copy notes at their seats	T – talks at blackboard Ss – listen and then work in their seats T – walks around the class	Some Ss work on the board T – discusses Ss' work on the board Ss – listen in their seats	T – talks at blackboard Ss – listen in their seats
Classroom talk	T – talks most of the time Pace relatively fast Convergent questions by T Conversation evaluation	T – talks most of the time Pace relatively slow Mostly convergent questions and some divergent questions Less evaluative	Some informal S talk (with each other) Pace relatively slow		
Content	Usually low demand of the cognitive processes	Higher demand in the cognitive processes Definitions/proofs/examples Heavy reliance on textbook	Medium demand on the cognitive processes Select exercises Focus on procedures or skills		
Climate	Serious Relatively quiet Mistakes less acceptable	Serious Relatively quiet Mistakes more acceptable	Less serious Less quiet Mistakes more acceptable		

SOURCE: U.S. Department of Education, National Center for Education Statistics, Third International Mathematics and Science Study (TIMSS), Video Study, 1999.

FIGURE E.4. Hypothesized country model for the Netherlands

Purpose	Re-instruction			Instruction		Assignment of task	Students attempt problems
Classroom routine	Going over old assignment *Nakijken*			Presenting new material		Assignment of task	Student problem solving continued work on old assignment and/or initial efforts on new assignment
Actions of participants	*Option 1* Completion of each problem as a class	*Option 2* T gives hints for selected problems		*Option 1* T verbalizes	*Option 2* Complete reliance on text	T – writes assignment on BB (or may give verbally)	If "Re-instruction" follows *Option 2*, Ss first work on the old assignment, then work on the new T – available to answer S-initiated questions T – gives mostly procedural assistance T – generally provides answers freely; doesn't require much S input T – may give semi-public assistance (at front of room) or private assistance (at Ss desks)
	T – goes through assignment, problem by problem at the front of the class, with or without use of the BB; Emphasis is on procedures	T – provides partial assistance (e.g., hints) on selected problems at the front of the class; T – provides answers on paper (e.g., answer sheet, access to T manual); Emphasis is on procedures		T – verbalizes text presentation and/or points to selected features of the text presentation	None		
	Ss – follow along at their desks, respond to T questions, ask clarifying questions	Ss – follow along at their desks; Very low S involvement		Ss – listen to T at their seats	Ss – read about new topic(s) from the text, at their desks	Ss – write assignment into their agendas	Ss – work in pairs at their desks and ask T for assistance when necessary, either at their desks or at the front of the room
Content	Small number of multi-part problems from the text (~5); Assignment given yesterday and worked on as HW; Generally one solution method provided			Heavy reliance on text; new material presented within the context of a task/problem		Small number of multi-part problems to be continued tonight as HW; Ss only need to find one solution method (any one solution is okay)	
	Problems are in a real-world context (might be considered "application"), situations vary across tasks, T rarely solicits errors						
Classroom talk	*Option 1* T asks Ss questions and rephrases Ss' responses	*Option 2* T briefly gives partial information on selected problems; Ss rarely ask questions Less S talk than in *Option 1*		*Option 1* Direct instruction	*Option 2* None	Direct instruction; T verbalizes the assignment as written on the BB	S-S talk regarding assignment; 1-on-1 (or 2- to 3-on-1) private, S-T conversations initiated by S, but then dominated by T
	Low level of evaluation/low concern for assessment						
Climate		High level of S freedom and responsibility			High level of S freedom and responsibility		Moderate level of noise is accepted by T
	High error tolerance by the T T-S relationship is relaxed						

SOURCE: U.S. Department of Education, National Center for Education Statistics, Third International Mathematics and Science Study (TIMSS), Video Study, 1999.

Hypothesized country models for Switzerland

FIGURE E.5. Hypothesized classroom patterns of Swiss mathematics lesson *with* introduction of new knowledge

Purpose	Opening	Construction of new cognitive structure *(Aufbau, Begriffsbildung)*	Working-through *(operatives Ueben)*	Practice (automatization, rehearsal) *Ueben*	
Classroom routine	Collecting homework, informal talk	Interactive instruction T – presentation 'real action' T – modeling problem solving	Interactive instruction Problem solving	Ss writing or reading at their desks Interactive instruction	
Actions of participants		T – asks questions and explains, demonstrates procedure, or states a problem… Ss – answer questions, observe T, imitate, act, solve problems; work as a whole class	(See Notes) Ss – work as a whole class	Ss – individual, group, or pair work	
Content		New concept is introduced in a step-by-step fashion, starting from Ss previous knowledge and/or their everyday experience Goal: Ss understand the concept (on their level of knowledge); Usefulness of concept (for further learning, and as a tool for everyday practice) emphasized; Visualization *(Anschauung)* is important; New information is reinforced (presented at board or textbook in a standardized fashion)	Sequence of carefully selected tasks related to new topic ("operatives *Ueben*")	Collection of tasks related to new topic	
Content (relationship between tasks)		Relationship between tasks: no set (often: problem-like situation)	Relationship between tasks: Set 2	Relationship between tasks: Set 1	Relationship between tasks: Set 1, 2, …
Classroom talk		*Lehrgespraech* (Interactive instruction; long wait-time, Ss expected to actively participate in construction process)	Interactive instruction	T-S-dialogue	
Climate		See Notes			

SOURCE: U.S. Department of Education, National Center for Education Statistics, Third International Mathematics and Science Study (TIMSS), Video Study, 1999.

210 Teaching Mathematics in Seven Countries
Results From the TIMSS 1999 Video Study

FIGURE E.6. Hypothesized classroom patterns of Swiss mathematics lessons *without* introduction of new knowledge

Purpose	Opening	Working-through or practice goal: understanding and/or proficiency	Practice goal: understanding and/or proficiency	Re-instruction, sharing	Practice goal: understanding and/or proficiency	Re-instruction, sharing	Using knowledge in different situations/to solve different problems *Anwenden*
Classroom routine	Collecting homework, informal talk	Interactive instruction	Ss solve tasks	Sharing and checking Ss' solutions (*Besprechung*) – interactive instruction – S presentation – discussion	Ss solve tasks	Sharing and checking Ss' solutions (*Besprechung*) – interactive instruction – S presentation – discussion	problem solving interactive instruction
Actions of participants		Classwork	Ss – individual, group, or pair work	Classwork	Ss – individual, group, or pair work	Classwork	Ss – individual, group, or pair work
Content		Topic: introduced in a previous lesson. T may start with short review of topic, and solving some examples of tasks			Progression to more demanding tasks, finally: to demanding application problems (possibly not in the same lesson, but later)		Character of tasks: Given new situations but connection to mathematical concepts is not obvious
Content (relationship between tasks)		Relationships between tasks: no set, or Set 1 or Set 2	Relationships between tasks: Set 1 or Set 2	Relationships between tasks: Set 1 or Set 2	Relationships between tasks: Set 1 or Set 2	Relationships between tasks: Set 1 or Set 2	Relationships between tasks: Set 2 or no set
Classroom talk		Interactive instruction	T-S-dialogue, and/or S-S-conversation	Interactive instruction/discussion	T-S-dialogue, and/or S-S-conversation	Interactive instruction/discussion	T-S-dialogue, S-S-conversation, discussion...
Climate							

SOURCE: U.S. Department of Education, National Center for Education Statistics, Third International Mathematics and Science Study (TIMSS), Video Study, 1999.

Notes to hypothesized classroom patterns of Swiss mathematics lessons with *introduction of new knowledge*

Most frequently a new topic (concept) might be co-constructed by means of interactive instruction *(Lehrgespraech)*. The means of guidance are primarily teacher questions and hints. The procedure is oriented toward the Socratic dialogue. The teacher questions serve two main purposes: (1) to guide and initiate students' thinking (e.g., propose a certain point of view, or perspective on a problem), and (2) to diagnose students' actual understanding. An important feature of quality of a *Lehrgespraech* is the need for sufficient wait-time after the teacher's questions.

The introduction phase may include some further actions that may be embedded in the interactive instruction, such as teacher presentation, or modeling or "real actions."

Reform 1:

In reform-oriented classrooms another pattern of introduction lessons might be expected: (1) student independent problem solving in pairs, groups, or individually (inventing procedures for solving new, open problems, discovering principles, regularities, and so on); (2) discussion of the different approaches and negotiating an accepted approach. This approach (influenced by scholars of mathematics didactics in Germany and the Netherlands) is presently recommended in teacher education and professional development. (It is unclear if this is observable at the eighth-grade level.)

Notes to hypothesized classroom patterns of Swiss mathematics lessons without *introduction of new knowledge*

As a general pattern an alternation between students solving tasks on their own and of sharing/checking/re-instruction based on students' work in a classwork sequence may be expected, but the duration of and total amount of the phases is not predictable.

The sequence of activity units varies, and does not always start with a classwork phase.

The first unit may provide some special kinds of tasks (warm-up, or a motivating starting task).

In most cases, the teacher will vary the social structure (e.g., classwork – individual work – classwork – pair work – and so on).

There is a progression from easier to more demanding tasks over the entire learning phase; usually the progression leads to application problems (most often, applied story problems).

Not all students always solve the same tasks (individualization of instruction).

Reform 2:

In some reform classrooms there will be no or almost no classwork phase and each student may be proceeding through a weekly assigned collection of learning tasks (arranged in collaboration with the teacher; individualized instruction). As with Reform 1, it is not clear if and how many teachers are in fact practicing this reform model of instruction (which is recommended in teacher development) at the eighth-grade level.

Teaching Mathematics in Seven Countries
Results From the TIMSS 1999 Video Study

FIGURE E.7. Hypothesized country model for the United States

Purpose	Review of previously learned material A			Acquisition of knowledge B	Practice and re-instruction C	
	Assess/evaluate	Assess/evaluate, re-instruct, secure knowledge	Secure knowledge, activate knowledge			
Classroom routine	Quiz A1	Checking homework A2	Warm-up/ brief review A3	Presenting new material B	Solving problems (not for homework OR for homework) C1	C2
Actions of participants	T – tells or solicits answers T – at the front	T – tells or solicits answers T – may work through difficult problems T – at the front	T – tells or solicits answers T – may work through problems T – at the front	Information provided mostly by T T – tells students when, why, and how to use certain procedures T – asks short-answer questions T – may do an example problem T – at the front	T – Ss through example problems T – at the front	T – walks around the room T – provides assistance to Ss who raise their hands
	Ss – Students take quiz Ss – provide or check their answers Ss – at their seats	Ss – provide or check their answers Ss – at their seats Ss – may put their answers on the board	Ss – complete problem(s) Ss – provide or check their answers Ss – at their seats	Ss – listen and answer T's questions Ss – may work on an activity, as explicitly instructed by the T Ss – at their seats	Ss – help the teacher do the problems Ss – at their seats	Ss – work individually or in small groups at their seats Ss – may state their answers as a class
Classroom talk	Known-answer questions, relatively quick pace, more student turns, T evaluates, recitation?	Known-answer questions, relatively quick pace, more student turns, T evaluates	Known-answer questions, relatively quick pace, more student turns, T evaluates	Fewer student turns, direction instruction? lecture?	Recitation, more student turns, direct instruction?	T-S dialogue, S-S dialogue (private talk)
Content	Content related to previous lesson	Content related to previous lesson	Content may or may not be closely related to the new topic	Simple rules or definitions stated by T, focus is mostly on procedure (little reflection on concepts)	More problems very similar to what the T has just shown	More problems very similar to what the T has just shown
Climate	T wants correct answers					Friendly atmosphere

SOURCE: U.S. Department of Education, National Center for Education Statistics, Third International Mathematics and Science Study (TIMSS), Video Study, 1999.

U.S. Notes

Recitation = A series of short, known-answer questions posed by the teacher, to solicit correct answers from students. Consists mainly of Initiation-Response-Evaluation sequences.

An alternative U.S. classroom pattern occasionally exists that does not resemble this model. These are considered "reform" mathematics lessons. They typically consist of an open-ended problem posed by the teacher, a long period of seatwork during which the students work on the problem, and then a period of "sharing" when the students provide their answers and the teacher summarizes the key points.

APPENDIX F
Numeric Values for the Lesson Signatures

TABLE F.1. Percentage of Australian lessons marked at each 10 percent interval of the lessons: 1999

	Beginning				Midpoint						End
	0	10	20	30	40	50	60	70	80	90	100
Review	87	77	52	37	32	27	23	23	22	23	23
Introduction of new content	12	23	33	40	36	31	33	33	29	23	23
Practice of new content	‡	‡	8	17	26	35	36	36	40	47	47
Public interaction	99	73	70	64	48	49	33	32	37	42	92
Private interaction	‡	36	37	40	52	54	67	72	66	59	9
Optional, student presents information	‡	‡	‡	‡	‡	‡	‡	‡	4	‡	‡
Mathematical organization	32	6	5	‡	‡	‡	‡	‡	‡	5	47
Non-problem	23	24	17	18	15	10	6	‡	7	10	36
Concurrent problem classwork	‡	14	15	10	7	8	5	6	14	17	‡
Concurrent problem seatwork	‡	31	26	30	40	43	54	62	62	56	6
Answered-only problems	‡	‡	‡	‡	‡	‡	‡	‡	‡	‡	‡
Independent problem 1	6	18	9	6	7	6	‡	5	4	‡	‡
Independent problem 2–5	‡	31	17	17	16	16	10	10	‡	5	‡
Independent problem 6–10	‡	21	23	6	10	7	10	10	5	‡	‡
Independent problems 11+	‡	‡	‡	8	8	9	8	10	5	6	‡

‡Reporting standards not met. Too few cases to be reported.

NOTE: The percentage of lessons coded for a feature at any point in time was calculated by dividing each lesson into 250 segments representing 0.4 percent of total lesson time. In a 50-minute lesson, this equates to segments of approximately 12 seconds each. Within each segment, the codes applied to the lessons are tabulated to derive the percentage of lessons exhibiting the feature. While many of the features listed above are mutually exclusive within each of the three dimensions (e.g., reviewing, introducing new content, and practicing new content within the purpose dimension), the percentages may not sum to 100 within a dimension due to the possibility of (a) a shift in codes within a segment in which case both codes would have been counted, (b) a segment being coded as "unable to make a judgment," (c) categories not reported, (d) momentary overlaps between the end of one feature and the start of another in which case both would be counted, and (e) rounding.
SOURCE: U.S. Department of Education, National Center for Education Statistics, Third International Mathematics and Science Study (TIMSS), Video Study, 1999.

TABLE F.2. Percentage of Czech lessons marked at each 10 percent interval of the lessons: 1999

	Beginning					Midpoint					End
	0	10	20	30	40	50	60	70	80	90	100
Review	99	96	52	79	62	50	40	35	30	30	32
Introduction of new content	‡	4	6	21	36	44	43	34	22	15	10
Practice of new content	‡	‡	‡	‡	‡	7	18	32	47	56	58
Public interaction	98	59	62	67	70	59	67	60	56	41	98
Private interaction	‡	30	28	20	8	19	13	20	23	41	‡
Optional, student presents information	‡	12	18	18	23	24	23	23	21	20	‡
Mathematical organization	17	‡	‡	‡	‡	‡	‡	‡	‡	‡	13
Non-problem	45	17	13	18	14	13	10	13	10	6	68
Concurrent problem classwork	‡	‡	11	12	14	11	9	9	6	5	5
Concurrent problem seatwork	‡	24	30	24	10	15	10	15	22	29	4
Answered-only problems	‡	‡	‡	‡	‡	‡	‡	‡	‡	‡	‡
Independent problem 1	7	26	18	11	10	5	6	‡	‡	‡	‡
Independent problem 2–5	‡	21	12	16	24	28	26	19	17	17	‡
Independent problem 6–10	‡	20	10	4	9	18	22	20	22	21	4
Independent problems 11+	‡	19	26	13	20	15	22	28	19	19	5

‡Reporting standards not met. Too few cases to be reported.
NOTE: The percentage of lessons coded for a feature at any point in time was calculated by dividing each lesson into 250 segments representing 0.4 percent of total lesson time. In a 50-minute lesson, this equates to segments of approximately 12 seconds each. Within each segment, the codes applied to the lessons are tabulated to derive the percentage of lessons exhibiting the feature. While many of the features listed above are mutually exclusive within each of the three dimensions (e.g., reviewing, introducing new content, and practicing new content within the purpose dimension), the percentages may not sum to 100 within a dimension due to the possibility of (a) a shift in codes within a segment in which case both codes would have been counted, (b) a segment being coded as "unable to make a judgment," (c) categories not reported, (d) momentary overlaps between the end of one feature and the start of another in which case both would be counted, and (e) rounding.
SOURCE: U.S. Department of Education, National Center for Education Statistics, Third International Mathematics and Science Study (TIMSS), Video Study, 1999.

218 | Teaching Mathematics in Seven Countries
Results From the TIMSS 1999 Video Study

TABLE F.3. Percentage of Hong Kong SAR lessons marked at each 10 percent interval of the lessons: 1999

	Beginning				Midpoint						End
	0	10	20	30	40	50	60	70	80	90	100
Review	77	58	33	26	22	14	13	12	12	10	8
Introduction of new content	23	40	58	55	47	40	40	40	29	23	23
Practice of new content	‡	‡	11	20	34	46	47	50	61	67	67
Public interaction	100	86	88	81	82	78	68	63	64	63	97
Private interaction	‡	11	10	19	19	20	26	31	28	31	4
Optional, student presents information	‡	4	3	‡	4	5	6	8	11	11	‡
Mathematical organization	16	‡	‡	‡	‡	‡	‡	‡	‡	3	15
Non-problem	43	29	23	12	11	8	9	4	‡	5	50
Concurrent problem classwork	‡	‡	‡	7	9	13	17	13	19	24	13
Concurrent problem seatwork	‡	6	10	15	21	22	23	30	32	32	3
Answered-only problems	‡	‡	‡	‡	‡	6	6	4	‡	‡	‡
Independent problem 1	20	38	27	10	7	6	6	4	‡	‡	‡
Independent problem 2–5	‡	27	36	44	42	34	23	21	17	16	‡
Independent problem 6–10	‡	‡	‡	8	11	14	15	21	14	11	‡
Independent problems 11+	‡	‡	4	‡	5	8	8	10	7	‡	4

‡Reporting standards not met. Too few cases to be reported.

NOTE: The percentage of lessons coded for a feature at any point in time was calculated by dividing each lesson into 250 segments representing 0.4 percent of total lesson time. In a 50-minute lesson, this equates to segments of approximately 12 seconds each. Within each segment, the codes applied to the lessons are tabulated to derive the percentage of lessons exhibiting the feature. While many of the features listed above are mutually exclusive within each of the three dimensions (e.g., reviewing, introducing new content, and practicing new content within the purpose dimension), the percentages may not sum to 100 within a dimension due to the possibility of (a) a shift in codes within a segment in which case both codes would have been counted, (b) a segment being coded as "unable to make a judgment," (c) categories not reported, (d) momentary overlaps between the end of one feature and the start of another in which case both would be counted, and (e) rounding.

SOURCE: U.S. Department of Education, National Center for Education Statistics, Third International Mathematics and Science Study (TIMSS), Video Study, 1999.

Appendix F
Numeric Values for the Lesson Signatures

TABLE F.4. Percentage of Japanese lessons marked at each 10 percent interval of the lessons: 1995

	Beginning					Midpoint					End
	0	10	20	30	40	50	60	70	80	90	100
Review	73	66	38	31	25	13	9	10	10	6	‡
Introduction of new content	27	38	62	72	73	75	71	62	58	55	55
Practice of new content	‡	‡	‡	‡	‡	12	20	28	33	39	40
Public interaction	98	64	53	67	67	59	65	35	57	62	98
Private interaction	‡	37	42	33	32	40	35	63	46	34	‡
Optional, student presents information	‡	‡	‡	‡	‡	‡	‡	8	‡	10	‡
Mathematical organization	‡	‡	‡	‡	‡	‡	‡	‡	‡	‡	16
Non-problem	31	32	21	17	15	12	17	‡	‡	15	55
Concurrent problem classwork	‡	‡	‡	9	‡	8	8	‡	‡	10	7
Concurrent problem seatwork	‡	9	10	‡	‡	‡	10	13	18	17	‡
Answered-only problems	‡	‡	‡	‡	‡	‡	‡	‡	‡	‡	‡
Independent problem 1	17	53	51	44	34	22	19	19	17	‡	‡
Independent problem 2–5	‡	‡	19	31	38	57	49	54	61	52	23
Independent problem 6–10	‡	‡	‡	‡	‡	‡	‡	‡	‡	‡	‡
Independent problems 11+	‡	‡	‡	‡	‡	‡	‡	‡	‡	‡	‡

‡Reporting standards not met. Too few cases to be reported.

NOTE: The percentage of lessons coded for a feature at any point in time was calculated by dividing each lesson into 250 segments representing 0.4 percent of total lesson time. In a 50-minute lesson, this equates to segments of approximately 12 seconds each. Within each segment, the codes applied to the lessons are tabulated to derive the percentage of lessons exhibiting the feature. While many of the features listed above are mutually exclusive within each of the three dimensions (e.g., reviewing, introducing new content, and practicing new content within the purpose dimension), the percentages may not sum to 100 within a dimension due to the possibility of (a) a shift in codes within a segment in which case both codes would have been counted, (b) a segment being coded as "unable to make a judgment," (c) categories not reported, (d) momentary overlaps between the end of one feature and the start of another in which case both would be counted, and (e) rounding.

SOURCE: U.S. Department of Education, National Center for Education Statistics, Third International Mathematics and Science Study (TIMSS), Video Study, 1999.

220 | Teaching Mathematics in Seven Countries
Results From the TIMSS 1999 Video Study

TABLE F.5. Percentage of Dutch lessons marked at each 10 percent interval of the lessons: 1999

	Beginning				Midpoint						End
	0	10	20	30	40	50	60	70	80	90	100
Review	64	61	53	46	36	30	26	25	25	24	24
Introduction of new content	29	31	36	37	37	34	34	29	28	27	27
Practice of new content	‡	‡	4	11	22	29	35	40	41	43	43
Public interaction	99	75	64	65	51	41	34	16	16	9	74
Private interaction	6	24	35	34	46	56	69	82	84	91	38
Optional, student presents information	‡	‡	‡	‡	‡	‡	‡	‡	‡	‡	‡
Mathematical organization	23	6	‡	‡	‡	‡	‡	‡	‡	‡	13
Non-problem	9	11	5	‡	6	‡	‡	‡	‡	‡	15
Concurrent problem classwork	‡	‡	‡	6	9	8	10	8	8	‡	4
Concurrent problem seatwork	‡	23	33	35	43	56	67	81	84	91	47
Answered-only problems	‡	7	8	‡	‡	‡	‡	‡	‡	‡	‡
Independent problem 1	16	24	9	7	‡	5	‡	‡	‡	‡	‡
Independent problem 2–5	‡	42	42	23	15	7	5	‡	‡	‡	‡
Independent problem 6–10	‡	19	13	32	21	12	7	6	‡	‡	‡
Independent problems 11+	‡	‡	8	9	10	14	10	‡	12	8	‡

‡Reporting standards not met. Too few cases to be reported.

NOTE: The percentage of lessons coded for a feature at any point in time was calculated by dividing each lesson into 250 segments representing 0.4 percent of total lesson time. In a 50-minute lesson, this equates to segments of approximately 12 seconds each. Within each segment, the codes applied to the lessons are tabulated to derive the percentage of lessons exhibiting the feature. While many of the features listed above are mutually exclusive within each of the three dimensions (e.g., reviewing, introducing new content, and practicing new content within the purpose dimension), the percentages may not sum to 100 within a dimension due to the possibility of (a) a shift in codes within a segment in which case both codes would have been counted, (b) a segment being coded as "unable to make a judgment," (c) categories not reported, (d) momentary overlaps between the end of one feature and the start of another in which case both would be counted, and (e) rounding.

SOURCE: U.S. Department of Education, National Center for Education Statistics, Third International Mathematics and Science Study (TIMSS), Video Study, 1999.

Appendix F
Numeric Values for the Lesson Signatures

TABLE F.6. Percentage of Swiss lessons marked at each 10 percent interval of the lessons: 1999

	Beginning				Midpoint						End
	0	10	20	30	40	50	60	70	80	90	100
Review	71	62	53	39	30	27	22	19	19	19	19
Introduction of new content	27	34	41	46	50	49	47	40	36	31	27
Practice of new content	‡	‡	7	14	19	20	28	37	42	48	51
Public interaction	98	73	67	57	51	48	44	44	41	45	93
Private interaction	3	29	33	39	46	49	58	55	58	61	13
Optional, student presents information	‡	‡	‡	‡	‡	‡	‡	‡	‡	‡	‡
Mathematical organization	21	‡	‡	‡	‡	‡	‡	‡	‡	‡	24
Non-problem	28	22	17	10	11	10	6	9	5	7	34
Concurrent problem classwork	‡	‡	9	11	8	10	12	14	14	18	5
Concurrent problem seatwork	3	23	27	32	36	41	49	48	52	55	13
Answered-only problems	2	6	3	‡	‡	3	‡	‡	‡	‡	‡
Independent problem 1	15	30	20	12	10	6	10	5	5	6	4
Independent problem 2–5	‡	18	23	21	21	20	14	12	11	13	3
Independent problem 6–10	‡	‡	13	5	5	4	4	4	5	‡	‡
Independent problems 11+	‡	‡	5	4	3	4	4	7	8	4	‡

‡Reporting standards not met. Too few cases to be reported.

NOTE: The percentage of lessons coded for a feature at any point in time was calculated by dividing each lesson into 250 segments representing 0.4 percent of total lesson time. In a 50-minute lesson, this equates to segments of approximately 12 seconds each. Within each segment, the codes applied to the lessons are tabulated to derive the percentage of lessons exhibiting the feature. While many of the features listed above are mutually exclusive within each of the three dimensions (e.g., reviewing, introducing new content, and practicing new content within the purpose dimension), the percentages may not sum to 100 within a dimension due to the possibility of (a) a shift in codes within a segment in which case both codes would have been counted, (b) a segment being coded as "unable to make a judgment," (c) categories not reported, (d) momentary overlaps between the end of one feature and the start of another in which case both would be counted, and (e) rounding.

SOURCE: U.S. Department of Education, National Center for Education Statistics, Third International Mathematics and Science Study (TIMSS), Video Study, 1999.

222 | Teaching Mathematics in Seven Countries
Results From the TIMSS 1999 Video Study

TABLE F.7. Percentage of U.S. lessons marked at each 10 percent interval of the lessons: 1999

	Beginning					Midpoint					End
	0	10	20	30	40	50	60	70	80	90	100
Review	90	87	79	62	50	45	40	35	34	30	30
Introduction of new content	10	14	22	33	31	38	21	25	20	19	17
Practice of new content	‡	‡	‡	7	19	22	40	41	48	50	51
Public interaction	76	74	79	81	75	68	63	67	59	45	79
Private interaction	26	28	22	19	25	34	36	34	41	55	28
Optional, student presents information	‡	‡	6	‡	‡	‡	‡	‡	‡	‡	‡
Mathematical organization	33	‡	‡	‡	‡	‡	‡	‡	‡	‡	29
Non-problem	15	21	13	19	11	9	4	8	5	9	25
Concurrent problem classwork	‡	10	17	14	5	‡	4	10	8	5	5
Concurrent problem seatwork	21	23	18	12	13	15	24	22	28	45	25
Answered-only problems	‡	4	10	5	10	‡	‡	‡	‡	‡	‡
Independent problem 1	8	30	18	15	13	13	7	7	7	6	‡
Independent problem 2–5	‡	11	37	44	30	24	16	11	5	8	‡
Independent problem 6–10	‡	‡	13	18	32	38	43	29	18	12	6
Independent problems 11+	‡	‡	‡	‡	‡	13	13	19	31	19	5

‡Reporting standards not met. Too few cases to be reported.

NOTE: The percentage of lessons coded for a feature at any point in time was calculated by dividing each lesson into 250 segments representing 0.4 percent of total lesson time. In a 50-minute lesson, this equates to segments of approximately 12 seconds each. Within each segment, the codes applied to the lessons are tabulated to derive the percentage of lessons exhibiting the feature. While many of the features listed above are mutually exclusive within each of the three dimensions (e.g., reviewing, introducing new content, and practicing new content within the purpose dimension), the percentages may not sum to 100 within a dimension due to the possibility of (a) a shift in codes within a segment in which case both codes would have been counted, (b) a segment being coded as "unable to make a judgment," (c) categories not reported, (d) momentary overlaps between the end of one feature and the start of another in which case both would be counted, and (e) rounding.

SOURCE: U.S. Department of Education, National Center for Education Statistics, Third International Mathematics and Science Study (TIMSS), Video Study, 1999.

ISBN 0-16-051381-2